T0351304

Advanced Structural Materials–2012

MATERIALS RESEARCH SOCIETY
SYMPOSIUM PROCEEDINGS VOLUME 1485

Advanced Structural Materials—2012

EDITORS

Hector A. Calderon
Escuela Superior de Física y Matemáticas
Instituto Politécnico Nacional
México D.F., México

Armando Salinas-Rodriguez
Centro de Investigación y de Estudios Avanzados
Instituto Politécnico Nacional
México D.F., México

Heberto Balmori-Ramirez
Escuela Superior de Ingeniería Química e
Industrias Extractivas
Instituto Politécnico Nacional
México D.F., México

Materials Research Society
Warrendale, Pennsylvania

Shaftesbury Road, Cambridge CB2 8EA, United Kingdom

One Liberty Plaza, 20th Floor, New York, NY 10006, USA

477 Williamstown Road, Port Melbourne, VIC 3207, Australia

314–321, 3rd Floor, Plot 3, Splendor Forum, Jasola District Centre, New Delhi – 110025, India

103 Penang Road, #05–06/07, Visioncrest Commercial, Singapore 238467

Cambridge University Press is part of Cambridge University Press & Assessment, a department of the University of Cambridge.

We share the University's mission to contribute to society through the pursuit of education, learning and research at the highest international levels of excellence.

www.cambridge.org
Information on this title: www.cambridge.org/9781605114620

Materials Research Society
506 Keystone Drive, Warrendale, PA 15086, USA
http://www.mrs.org

First published 2013

CODEN: MRSPDH

A catalogue record for this publication is available from the British Library

ISBN 978-1-605-11462-0 Hardback

CONTENTS

Preface ix

Materials Research Society Symposium Proceedings xi

Prediction of Hot Flow Curves of Construction Steels by Physically-Based Constitutive Equations 1
G. Varela-Castro and J.M. Cabrera

Recent Advances on Bulk Tantalum Carbide Produced by Solvothermal Synthesis and Spark Plasma Sintering 9
Braeden M. Clark, James P. Kelly, and Olivia A. Graeve

Removal of Lead Ions in an Aqueous Medium with Activated Carbon Fibers 21
Teresa Ramírez-Rodríguez, and
Fray de Landa Castillo-Alvarado

Finite Element Simulation of Steel Quench Distortion- Parametric Analysis of Processing Variables 29
F.A. García-Pastor, R.D. López-García, E. Alfaro-López,
and M.J. Castro-Román

3D Computational Simulation of Multi-Impact Shot Peening 35
Juan Solórzano-López, and Francisco Alfredo García-Pastor

Study of Aluminum Degasification with Impeller-injector Assisted by Physical Modeling 41
M. Hernández-Hernández, E.A. Ramos-Gómez,
and M.A. Ramírez-Argáez

Glass Ceramic Materials of the SiO_2-CaO-MgO-Al_2O_3 System: Structural Characterization and Fluorine Effect 47
Mitzué Garza-García, Jorge López-Cuevas,
and Oscar Hernández-Ibarra

Effect of Controlled Corrosion Attack with HCl Acid on the Fatigue Endurance of Aluminum Alloy AISI 6063-T5, under Rotating Bending Fatigue Tests 53
G.M. Domínguez-Almaraz, J.L. Ávila-Ambriz, F. Peyraut,
and E. Cadenas-Calderón

The Influence of Modifiers on the Pigmentary Properties of Titanium Dioxide 59
Marta A. Gleń, and Barbara U. Grzmil

Phase Formation at Selective Laser Synthesis in Al_2O_3–TiO_2-Y_2O_3 Powder Compositions 65
M. Vlasova, M. Kakazey, P.A. Márquez-Aguilar, A. Ragulya, V. Stetsenko, and A. Bykov

Nanostructured Ceramic Oxides Containing Ferrite Nanoparticles and Produced by Mechanical Milling 71
A. Huerta-Ricardo, K. Tsuchiya, T. Umemoto, and H.A. Calderon

Recycled HDPE-Tetrapack Composites. Isothermal Crystallization, Light Scattering and Mechanical Properties 77
A. Parada-Soria, H.F. Yao, B. Alvarado-Tenorio, L. Sanchez-Cadena, and A. Romo-Uribe

Effect of Boron on the Continuous Cooling Transformation Kinetics in a Low Carbon Advanced Ultra-high Strength Steel (A-UHSS) 83
G. Altamirano, I. Mejía, A. Hernández-Expósito, and J.M. Cabrera

Effect of Mechanical Activation on the Crystallization and Properties of Iron-rich Glass Materials 89
Claudia M. García-Hernández, Jorge López-Cuevas, and José L. Rodríguez-Galicia

Effects of Holding Time on HAZ-Softening in Resistance Spot Welded DP980 Steels 95
C.J. Martínez-González, A. López-Ibarra, S. Haro-Rodriguez, V.H. Baltazar-Hernandez, S.S. Nayak, and Y. Zhou

Effect of Slag on Mixing Time in Gas-stirred Ladles Assisted with a Physical Model 101
Adrián M. Amaro-Villeda, A. Conejo, and Marco A. Ramírez-Argáez

Synthesis and Characterization of Ceramic Composites of the Binary System $Ba_{0.75}Sr_{0.25}AlSi_2O_8$ - Al_2O_3 107
Jorge López-Cuevas, Magaly V. Ramos-Ramírez, and José L. Rodríguez-Galicia

Effect of the Ratio Mo/Cr in the Precipitation and Distribution of Carbides in Alloyed Nodular Iron 113
H.D. Rivero, José A. García, E. Cándido Atlatenco, Alejandro D. Basso, and J. Sicora

Decarburization of Hot-rolled Non-oriented Electrical Steels 119
Emmanuel J. Gutiérrez, and Armando Salinas

Effect of Cooling Rate on the Formation and Distribution of Carbides in Nodular Iron Alloyed with Cr 125
Ramses Zenil, Jose A. García, Gerardo A. Ruiz, Alejandro D. Basso, and Jorge. Sicora

Effects of Austenitizing Temperature and Cooling Rate on the Phase Transformation Texture in Hot Rolled Steels 131
N.M. López, and A. Salinas R

Fracture Behavior of Heat Treated Liquid Crystalline Polymers 137
A. Reyes-Mayer, B. Alvarado-Tenorio, A. Romo-Uribe, O. Flores, B. Campillo, and M. Jaffe

Dynamically Recrystallized Austenitic Grain in a Low Carbon Advanced Ultra-high Strength Steel (A-UHSS) Microalloyed with Boron under Hot Deformation Conditions 143
I. Mejía, E. García-Mora, G. Altamirano, A. Bedolla-Jacuinde, and J.M. Cabrera

Magnetization Study of the Kinetic Arrest of Martensitic Transformation in As-quenched $Ni_{52.2}Mn_{34.3}In_{13.5}$ Melt Spun Ribbons 149
F.M. Lino-Zapata, J.L. Sánchez Llamazares, D. Ríos-Jara, A.G. Lara-Rodríguez, and T. García-Fernández

Thermal Properties of Cu-Hf-Ti Metallic Glass Compositions 155
I.A. Figueroa

On the Characterization of Eutectic Grain Growth during Solidification 161
M. Morua, M. Ramirez-Argaez, and C. Gonzalez-Rivera

Author Index 167

Subject Index 169

PREFACE

This volume is a compilation of papers that were presented at the Symposium 7D of the XXI International Conference of the Mexican Academy of Materials Science – MRS Mexico, organized in collaboration with MRS, "Advanced Structural Materials" which was held from August 13th to 18th, 2012 in Cancun, Mexico. The symposium was devoted to fundamental and technological applications of structural materials, and continued the tradition of providing a forum for scientists from various backgrounds with a common interest in the development and use of structural materials to come together and share their findings and expertise.

The papers contained in this volume are a collection of invited and contributed papers. This year, the symposium was attended by participants from Argentina, Brazil, Canada, Czech Republic, France, India, Japan, Korea, Mexico, Poland, Turkey, Spain, Switzerland, The United States, Ukraine and United Kingdom. All papers have been thoroughly reviewed by at least two referees and edited to the standards of the Materials Research Society. We are grateful to all those referees who, by their comments and constructive criticism, helped to improve the finally printed papers, and to all the authors who made additional efforts to prepare their manuscripts.

The Advanced Structural Materials symposium has been held for the last 15 years with the objective of presenting overviews and recent investigations related to advanced structural metallic, ceramic and composite materials. The topics include innovative processing, phase transformations, mechanical properties, oxidation resistance, modeling and the relationship between processing, microstructure and mechanical behavior. Additionally, papers on industrial application and metrology are included.

The organizing committee gratefully acknowledges the enthusiastic cooperation of all symposium participants, as well as the kind acceptance of the editorial committee of Materials Research Society to publish these proceedings. The financial support of the Instituto Politécnico Nacional (IPN, Mexico) and the Centro de Investigación y de Estudios Avanzados del Instituto Politécnico Nacional (CINVESTAV-IPN) is also acknowledged. We hope that all readers will come to consider Advanced Structural Materials symposium at Cancun as a suitable forum to present results of their recent research and experience.

Greetings from the Organizing Committee

The organizing committee:

Heberto Balmori-Ramírez (Instituto Politécnico Nacional).
Hector A. Calderon-Benavides (Instituto Politécnico Nacional).
Armando Salinas-Rodríguez (Centro de Investigación y de Estudios Avanzados del IPN).

MATERIALS RESEARCH SOCIETY SYMPOSIUM PROCEEDINGS

Volume 1477 — Low-Dimensional Bismuth-based Materials, 2012, S. Muhl, R. Serna, A. Zeinert, S. Hirsekor, ISBN 978-1-60511-454-5
Volume 1478 — Nanostructured Carbon Materials for MEMS/NEMS and Nanoelectronics, 2012, A.V. Sumant, A.A. Balandin, S.A. Getty, F. Piazza, ISBN 978-1-60511-455-2
Volume 1479 — Nanostructured Materials and Nanotechnology—2012, 2012, C. Gutiérrez-Wing, J.L. Rodríguez-López, O. Graève, M. Munoz-Navia, ISBN 978-1-60511-456-9
Volume 1480 — Novel Characterization Methods for Biological Systems, 2012, P.S. Bermudez, J. Majewski, N. Alcantar, A.J. Hurd, ISBN 978-1-60511-457-6
Volume 1481 — Structural and Chemical Characterization of Metals, Alloys and Compounds—2012, 2012, A. Contreras Cuevas, R. Pérez Campos, R. Esparza Muñoz, ISBN 978-1-60511-458-3
Volume 1482 — Photocatalytic and Photoelectrochemical Nanomaterials for Sustainable Energy, 2012, L. Guo, S.S. Mao, G. Lu, ISBN 978-1-60511-459-0
Volume 1483 — New Trends in Polymer Chemistry and Characterization, 2012, L. Fomina, M.P. Carreón Castro, G. Cedillo Valverde, J. Godínez Sánchez, ISBN 978-1-60511-460-6
Volume 1484 — Advances in Computational Materials Science, 2012, E. Martínez Guerra, J.U. Reveles, A. Aguayo González, ISBN 978-1-60511-461-3
Volume 1485 — Advanced Structural Materials—2012, 2012, H. Calderon, H.A. Balmori, A. Salinas, ISBN 978-1-60511-462-0
Volume 1486E — Nanotechnology-enhanced Biomaterials and Biomedical Devices, 2012, L. Yang, M. Su, D. Cortes, Y. Li, ISBN 978-1-60511-463-7
Volume 1487E — Biomaterials for Medical Applications—2012, 2012, S. Rodil, A. Almaguer, K. Anselme, J. Castro, ISBN 978-1-60511-464-4
Volume 1488E — Concrete with Smart Additives and Supplementary Cementitious Materials, 2012, L.E. Rendon Diaz Miron, B. Martinez Sanchez, K. Kovler, N. De Belie, ISBN 978-1-60511-465-1
Volume 1489E — Compliant Energy Sources, 2013, D. Mitlin, ISBN 978-1-60511-466-8
Volume 1490 — Thermoelectric Materials Research and Device Development for Power Conversion and Refrigeration, 2013, G.S. Nolas, Y. Grin, A. Thompson, D. Johnson, ISBN 978-1-60511-467-5
Volume 1491E — Electrocatalysis and Interfacial Electrochemistry for Energy Conversion and Storage, 2013, T.J. Schmidt, V. Stamenkovic, M. Arenz, S. Mitsushima, ISBN 978-1-60511-468-2
Volume 1492 — Materials for Sustainable Development—Challenges and Opportunities, 2013, M-I. Baraton, S. Duclos, L. Espinal, A. King, S.S. Mao, J. Poate, M.M. Poulton, E. Traversa, ISBN 978-1-60511-469-9
Volume 1493 — Photovoltaic Technologies, Devices and Systems Based on Inorganic Materials, Small Organic Molecules and Hybrids, 2013, K.A. Sablon, J. Heier, S.R. Tatavarti, L. Fu, F.A. Nuesch, C.J. Brabec, B. Kippelen, Z. Wang, D.C. Olson, ISBN 978-1-60511-470-5
Volume 1494 — Oxide Semiconductors and Thin Films, 2013, A. Schleife, M. Allen, S.M. Durbin, T. Veal, C.W. Schneider, C.B.Arnold, N. Pryds, ISBN 978-1-60511-471-2
Volume 1495E — Functional Materials for Solid Oxide Fuel Cells, 2013, J.A. Kilner, J. Janek, B. Yildiz, T. Ishihara, ISBN 978-1-60511-472-9
Volume 1496E — Materials Aspects of Advanced Lithium Batteries, 2013, V. Thangadurai, ISBN 978-1-60511-473-6
Volume 1497E — Hierarchically Structured Materials for Energy Conversion and Storage, 2013, P.V. Braun, ISBN 978-1-60511-474-3
Volume 1498 — Biomimetic, Bio-inspired and Self-Assembled Materials for Engineered Surfaces and Applications, 2013, M.L. Oyen, S.R. Peyton, G.E. Stein, ISBN 978-1-60511-475-0
Volume 1499E — Precision Polymer Materials—Fabricating Functional Assemblies, Surfaces, Interfaces and Devices, 2013, C. Hire, ISBN 978-1-60511-476-7
Volume 1500E — Next-Generation Polymer-Based Organic Photovoltaics, 2013, M.D. Barnes, ISBN 978-1-60511-477-4
Volume 1501E — Single-Crystalline Organic and Polymer Semiconductors—Fundamentals and Devices, 2013, S.R. Parkin, ISBN 978-1-60511-478-1
Volume 1502E — Membrane Material Platforms and Concepts for Energy, Environment and Medical Applications, 2013, B. Hinds, F. Fornasiero, P. Miele, M. Kozlov, ISBN 978-1-60511-479-8
Volume 1503E — Colloidal Crystals, Quasicrystals, Assemblies, Jammings and Packings, 2013, S.C. Glotzer, F. Stellacci, A. Tkachenko, ISBN 978-1-60511-480-4
Volume 1504E — Geometry and Topology of Biomolecular and Functional Nanomaterials, 2013, A. Saxena, S. Gupta, R. Lipowsky, S.T. Hyde, ISBN 978-1-60511-481-1

MATERIALS RESEARCH SOCIETY SYMPOSIUM PROCEEDINGS

Volume 1505E — Carbon Nanomaterials, 2013, J.J. Boeckl, W. Choi, K.K.K. Koziol, Y.H. Lee, W.J. Ready, ISBN 78-1-60511-482-8
Volume 1506E — Combustion Synthesis of Functional Nanomaterials, 2013, R.L. Vander Wal, ISBN 978-1-60511-483-5
Volume 1507E — Oxide Nanoelectronics and Multifunctional Dielectrics, 2013, P. Maksymovych, J.M. Rondinelli, A. Weidenkaff, C-H. Yang, ISBN 978-1-60511-484-2
Volume 1508E — Recent Advances in Optical, Acoustic and Other Emerging Metamaterials, 2013, K. Bertoldi, N. Fang, D. Neshev, R. Oulton, ISBN 978-1-60511-485-9
Volume 1509E — Optically Active Nanostructures, 2013, M. Moskovits, ISBN 978-1-60511-486-6
Volume 1510E — Group IV Semiconductor Nanostructures and Applications, 2013, L. Dal Negro, C. Bonafos, P. Fauchet, S. Fukatsu, T. van Buuren, ISBN 978-1-60511-487-3
Volume 1511E — Diamond Electronics and Biotechnology—Fundamentals to Applications VI, 2013, Y. Zhou, ISBN 978-1-60511-488-0
Volume 1512E — Semiconductor Nanowires—Optical and Electronic Characterization and Applications, 2013, J. Arbiol, P.S. Lee, J. Piqueras, D.J. Sirbuly, ISBN 978-1-60511-489-7
Volume 1513E — Mechanical Behavior of Metallic Nanostructured Materials, 2013, Q.Z. Li, D. Farkas, P.K. Liaw, B. Boyce, J.Wang, ISBN 978-1-60511-490-3
Volume 1514 — Advances in Materials for Nuclear Energy, 2013, C.S. Deo, G. Baldinozzi, M.J. Caturia, C-C. Fu, K. Yasuda, Y. Zhang, ISBN 978-1-60511-491-0
Volume 1515E — Atomic Structure and Chemistry of Domain Interfaces and Grain Boundaries, 2013, S.B. Sinnott, B.P. Uberuaga, E.C. Dickey, R.A. De Souza, ISBN 978-1-60511-492-7
Volume 1516 — Intermetallic-Based Alloys—Science, Technology and Applications, 2013, I. Baker, S. Kumar, M. Heilmaier, K. Yoshimi, ISBN 978-1-60511-493-4
Volume 1517 — Complex Metallic Alloys, 2013, M. Feuerbacher, Y. Ishii, C. Jenks, V. Fournée, ISBN 978-1-60511-494-1
Volume 1518 — Scientific Basis for Nuclear Waste Management XXXVI, 2012, N. Hyatt, K.M. Fox, K. Idemitsu, C. Poinssot, K.R. Whittle, ISBN 978-1-60511-495-8
Volume 1519E — Materials under Extreme Environments, 2013, R.E. Rudd, ISBN 978-1-60511-496-5
Volume 1520E — Structure-Property Relations in Amorphous Solids, 2013, Y. Shi, M.J. Demkowicz, A.L. Greer, D. Louca, ISBN 978-1-60511-497-2
Volume 1521E — Properties, Processing and Applications of Reactive Materials, 2013, E. Dreizin, ISBN 978-1-60511-498-9
Volume 1522E — Frontiers of Chemical Imaging—Integrating Electrons, Photons and Ions, 2013, C.M. Wang, J.Y. Howe, A. Braun, J.G. Zhou, ISBN 978-1-60511-499-6
Volume 1523E — Materials Informatics, 2013, R. Ramprasad, R. Devanathan, C. Breneman, A. Tkatchenko, ISBN 978-1-60511-500-9
Volume 1524E — Advanced Multiscale Materials Simulation—Toward Inverse Materials Computation, 2013, D. Porter, ISBN 978-1-60511-501-6
Volume 1525E — Quantitative *In-Situ* Electron Microscopy, 2013, N.D. Browning, ISBN 978-1-60511-502-3
Volume 1526 — Defects and Microstructure Complexity in Materials, 2013, A. El-Azab, A. Caro, F. Gao, T. Yoshiie, P. Derlet, ISBN 978-1-60511-503-0
Volume 1527E — Scanning Probe Microscopy—Frontiers in Nanotechnology, 2013, M. Rafailovich, ISBN 978-1-60511-504-7
Volume 1528E — Advanced Materials Exploration with Neutrons and Synchrotron X-Rays, 2013, J.D. Brock, ISBN 978-1-60511-505-4
Volume 1529E — Roll-to-Roll Processing of Electronics and Advanced Functionalities, 2013, T. Blaudeck, G. Cho, M.R. Dokmeci, A.B. Kaul, M.D. Poliks, ISBN 978-1-60511-506-1
Volume 1530E — Materials and Concepts for Biomedical Sensing, 2013, P. Kiesel, M. Zillmann, H. Schmidt, B. Hutchison, ISBN 978-1-60511-507-8
Volume 1531E — Low-Voltage Electron Microscopy and Spectroscopy for Materials Characterization, 2013, R.F. Egerton, ISBN 978-1-60511-508-5
Volume 1532E — Communicating Social Relevancy in Materials Science and Engineering Education, 2013, K. Chen, R. Nanjundaswamy, A. Ramirez, ISBN 978-1-60511-509-2
Volume 1533E — The Business of Nanotechnology IV, 2013, L. Merhari, D. Cruikshank, K. Derbyshire, J. Wang, ISBN 978-1-60511-510-8

Materials Research Society Symposium Proceedings

Volume 1534E — Low-Dimensional Semiconductor Structures, 2012, T. Torchyn, Y. Vorobie, Z. Horvath, ISBN 978-1-60511-511-5

Prior Materials Research Society Symposium Proceedings available by contacting Materials Research Society

Mater. Res. Soc. Symp. Proc. Vol. 1485 © 2013 Materials Research Society
DOI: 10.1557/opl.2013.206

Prediction of hot flow curves of construction steels by physically-based constitutive equations

G.Varela-Castro[1], J.M. Cabrera[1,2]
[1]Fundació CTM Centre Tecnològic, Av. de les Bases de Manresa 1, 08242-Manresa, Spain.
[2]Departament de Ciència del Materials i Enginyeria Metal·lúrgica, ETSEIB, Universitat Politècnica de Catalunya, Av. Diagonal 647, 08028-Barcelona, Spain.
jose.maria.cabrera@upc.edu.

ABSTRACT

The development of accurate constitutive equations is important for the success of computer simulations of high temperature forming operations. Often, these simulations must be made on alloys that have not been completely characterized. For that reason physically-based constitutive equations taking the chemical composition into consideration, involving deformation mechanisms and characteristic properties of the material are necessary. The influence that exerts the solute elements to an alloy on the mechanisms of diffusion on deformation processes at high temperatures is not an easy subject and the available information in literature is scarce.

This study examines that influence working on the basis of eight structural plain carbon steels with the chemical composition ranging between 0.15-0.45%C, 0.2-0.4%Si and 0.6-1.6%Mn produced by Electro-Slag Remelting ESR process and tested by isothermal uniaxial compression technique. The studied deformation conditions include strain rates ranging between $5\cdot10^{-4}$ to $1\cdot10^{-1}$ s^{-1} and temperatures between 0.6-0.75T_m, with T_m the melting temperature.

A constitutive expression for the hot working behavior is proposed, it includes the variation of the diffusion parameters with the chemical composition. To such aim the effect of the chemical composition of the alloy on the pre-exponential factor D_0 of the gamma iron self-diffusion coefficient D_{sd} is included. Finally, a comparison of the experimental and predicted results shows the good agreement of the model with experimental flow data.

INTRODUCTION

Accurate constitutive equations are fundamental in order to success in hot working simulations, and the availability of reliable material data is crucial in designing materials and manufacturing processes [1]. In a given hot deformation process the diffusion kinetics are clearly influenced by the solute elements into an alloy. Excellent references about diffusion and self-diffusion for different kinds of metals and alloys can be found [2,3]. However it is easy to see that there is relatively little information about the study of the effect of solute elements, in combination, on the self-diffusion coefficient of Fe in austenite (γ-Fe), in particular on the pre-exponential factor of diffusion coefficient, D_0. This factor can be considered as constant, nevertheless there is experimental evidence that reveals a variation of D_0 with chemical composition. Traditionally the diffusion of C into austenite, for example, has received much attention due to its importance in the design and application of many heat treatments of steels. Therefore it is possible to find references showing different types of variation of the pre-exponential factor with C concentration [4,5]. However, the problem difficulty increases as the steels become more alloyed; it means multicomponent systems [6-8].

Here it is assumed that, in the modeling of a particular thermomechanical process, the influence of chemical composition is revealed through the frequency factor D_0 of the self-diffusion coefficient of Fe in γ-Fe. Applying the modified hyperbolic sine equation [9], the effects of carbon, silicon and manganese on the frequency factor are discussed.

EXPERIMENT

This study examines the influence of the chemical composition on the diffusion behavior on the basis of structural plain carbon steels produced by Electro-Slag Remelting process and tested by isothermal uniaxial compression technique. The chemical composition of the steels is detailed in Table 1.

Table 1. Chemical composition of alloys examined, mass %. (P: ~0.03 and S: ~0.01).

	A000	**A001**	**A010**	**A011**	**A100**	**A101**	**A110**	**A111**
C	0.160	0.160	0.150	0.180	0.450	0.460	0.450	0.470
Si	0.250	0.250	0.450	0.470	0.240	0.240	0.450	0.460
Mn	0.670	1.700	0.630	1.660	0.650	1.730	0.730	1.650
Cr	0.180	0.170	0.180	0.075	0.220	0.220	0.220	0.220
Mo	0.032	0.032	0.034	0.025	0.010	0.010	0.010	0.010
Al	0.009	0.007	0.011	0.007	0.008	0.008	0.012	0.011
Cu	0.250	0.240	0.250	0.230	0.010	0.020	0.020	0.020
Fe	Bal.	Bal.	Bal.	Bal.	Bal.	Bal.	Bal.	Bal.

A universal compression testing machine is employed to carry out the mentioned tests. Cylindrical specimens are machined with 8 mm of diameter and a diameter/length ratio of 1.5. See Fig. 1.

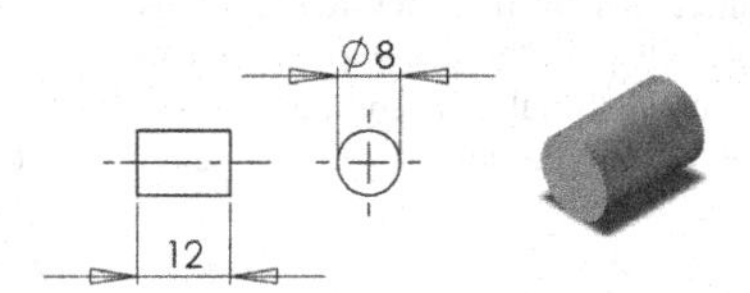

Fig. 1. Schematic of compression samples.

Tests are performed at temperatures ranging between 900 and 1100 °C and four different strain rates are applied at each temperature, namely $5\cdot10^{-4}$, $1\cdot10^{-3}$, $1\cdot10^{-2}$ and $1\cdot10^{-1}$ s^{-1}. The applied thermal cycle comprising: (i) reheating at 2 °Cs^{-1} until 1100°C, (ii) holding time of 300 s at this temperature, (iii) samples are then cooled at 2 °Cs^{-1} to the test temperature and (iv) finally a holding time period of 300 s before testing. The maximum applied true strain is 0.8. Specimens tested at 1100°C are immediately deformed after soaking. All tests are performed under Ar atmosphere and, in order to prevent friction, Ta foils with B nitride powder are used. The initial grain sizes obtained at the solubilizing temperature remain approximately constant for the all alloys (70 to 100 μm). Therefore, it is considered analogous for all alloys and no influence on hot flow behavior is expected.

DISCUSSION

In a general manner the flow stress-strain (σ-ε) curves are classical in morphology and representative of the occurrence of dynamic recrystallization DRX during a given thermomechanical process. See Fig. 2 and Fig. 3 (a,b). σ-ε curves consisting of an initial work hardening zone, at lower strains, followed by a work softening zone and a peak stress at higher strains [10]. Nevertheless at low temperatures and/or high strain rates, the only softening mechanism involved is dynamic recovery, DRV (See Fig. 2(d)). At low strain rates and relatively high temperatures, multiple peak DRX behavior is observed. Bergström [11,12] and Kocks-Mecking [13,14] approaches are utilized to derive the flow stress behavior of the steels and KJMA [15-18] approach to derive the DRX kinetics respectively, as reported by Cabrera et al. [19]. The Fig. 2(a-d) also shows, in dashed line, the modeled curves using the mentioned models for the alloy A000.

In order to analyze the influence of chemical composition (C, Si and Mn) on the pre-exponential factor of the self-diffusion coefficient of Fe in austenite, the data collected from the experimental curves (160) are treated using the modified hyperbolic sine equation reported by Frost and Ashby [9], representing the strain rate (s^{-1}) as follow:

$$\dot{\varepsilon} = A'\left(\frac{D_{eff}Gb}{kT}\right)\left[\sinh\left(\alpha\frac{\sigma_i}{G}\right)\right]^n \quad (1)$$

where A' and α are constants, k the Boltzmann constant (JK^{-1}), D_{eff} the effective diffusion coefficient (m^2s^{-1}), G the shear modulus (MPa) [9], b the Burgers vector (m), n the exponent of creep equal to 5, T the absolute temperature (K) and σ_i the characteristic stress (MPa); in this case the maximum stress or the peak stress σ_p.

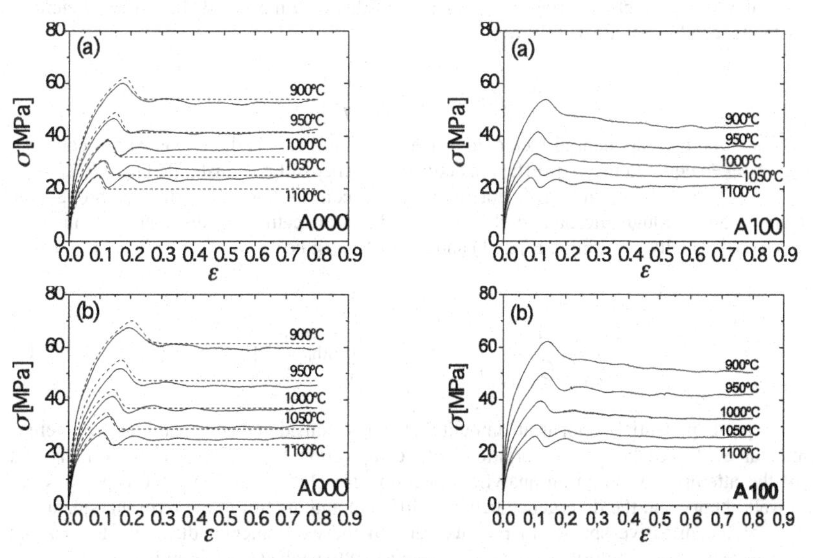

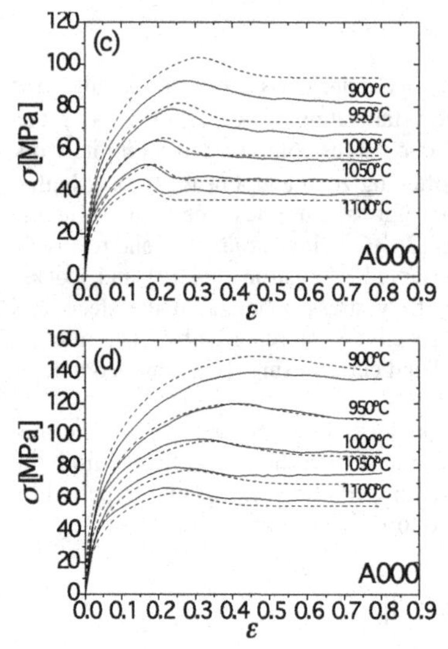

Fig. 2. Flow curves for A000 alloy at: (a) $5\cdot10^{-4}$, (b) $1\cdot10^{-3}$, (c) $1\cdot10^{-2}$, (d) $1\cdot10^{-1}$ s^{-1}. Dashed lines represent modeled stress-strain curves.

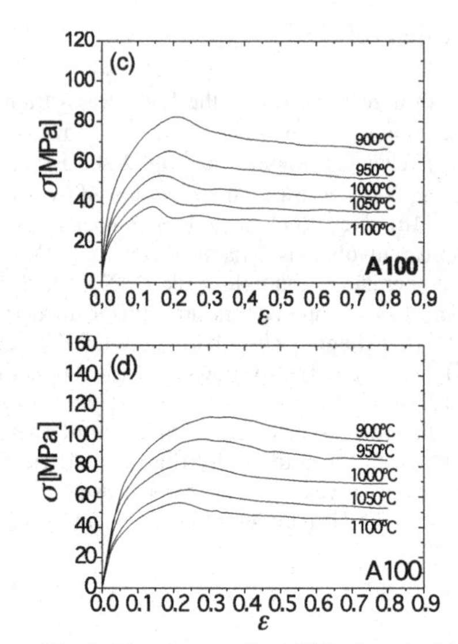

Fig. 3. Flow curves for A100 alloy at: (a) $5\cdot10^{-4}$, (b) $1\cdot10^{-3}$, (c) $1\cdot10^{-2}$, (d) $1\cdot10^{-1}$ s^{-1}.

Moreover the effective diffusion coefficient, in this case the self-diffusion coefficient of Fe in austenite, can be defined as follows:

$$D_{sd\_eff} = D_{0\_eff} \exp\left(\frac{-Q_{eff}}{RT}\right) \tag{2}$$

where D_{0_eff} is the pre-exponential factor (m^2s^{-1}) and R ($Jmol^{-1}K^{-1}$) is the universal gas constant. Here, Q_{eff} is taken equal to the self-diffusion activation energy of Fe in austenite Q_{sd}; 270 $kJmol^{-1}$ [9,19]. The effective diffusion coefficient involves the contributions of diffusion through the crystal lattice and through the named short circuit diffusion paths, i.e. dislocations and grain boundaries [20]. Finally combining Eqn. (1) and Eqn. (2) must be:

$$\left(\frac{\dot{\varepsilon}kT}{\exp\left(\frac{-Q_{sd}}{RT}\right)Gb}\right)^{\frac{1}{n}} = D_{0-eff}^{1/n} A'^{1/n} \cdot \left(\sinh\left(\frac{\alpha\sigma_p}{G}\right)\right) \tag{3}$$

The pre-exponential factor is an amount that can be considered constant, however there is experimental evidence showing a variation with composition [2,3]. Zener [21] explains that many of the attempts to obtain an analytic expression for D_{0_eff} begin with the application of random walk theory and the divergence between different authors lies in the term that quantifies jump rate of the diffusive specie. In practice, due to the experimental difficulty to measure diffusion, more pragmatic solutions are used, i.e. statistically analysis of literature data [4].

Some analytical dependencies of the diffusion coefficient of C in austenite as a function of the C content, where exponential and polynomial type relationships for D_{0_eff} have been reported [5]. More recently Lee et al. [4] propose an expression for the diffusivity of C in austenite, not only as a function of C content, also as a function of the alloying elements, whether interstitial or substitutional ones. Mead and Birchenall [27] studied the self-diffusion of Fe in austenite as a function of C concentration in Fe-C alloys up to 0-1.4%C. The results indicate that while the diffusivity systematically increases with an increasing content of C, the activation energy decreases. At the same time it is observed that the value of D_{0_eff} decreases appreciably with the content of C. Treheux et al. [28] studied the effect of Si content on the self-diffusion of Fe in Fe-Si alloys. For Si content less than 1% there is slight decrease on self-diffusivities and an increase of activation energy. Finally, in the case of Mn, and according to reported values [19] there is a slight or no influence of Mn content on the D_{0_eff} factor, for similar contents to those studied here. For much higher Mn values, D_{0_eff} grows approximately linearly with Mn concentration.

The values of the pre-exponential factor are calculated replacing $D_{0\_eff}^{1/n}$ in the Eqn. (3) by the following expression:

$$D_{0\_eff}^{1/n} = a_1 + a_2\%C^{a_3} + a_4\%Si + a_5\%Mn \qquad (4)$$

where a_1 to a_5 are constants and %C, %Si and %Mn represent the mass content of carbon, manganese and silicon respectively. A diversity of non-linear regression equations for $D_{0\_eff}^{1/n}$ where tried. Betters results are obtained using the previous expression with a regression coefficient of R^2=0.98. Table 2 shows those values and Table 3 shows the D_{0_eff} values calculated from values in Table 2 for all the alloys tested in the rage of temperature of 900 to 1100°C.

Table 2. Fit parameters.

a_1	a_2	a_3	a_4	a_5	$(A')^{1/n}$	α
0.1125	0.1028	2.1420	-0.0249	-0.0079	0.0547	469

where a_1^n, with n=5, is equal to the pre-exponential factor for self-diffusion of pure Fe (1.8E-05 m^2s^{-1}) [9,24]. According to Table 2, C has the greatest influence on the value of D_{0_eff} meanwhile Si and Mn present lower factors and even the latter is one order of magnitude lower than silicon. Values of A' and α are taken as "universals" for this family of alloys. Note that in this approach only chemical composition and deformation conditions (strain rate and temperature) have been used. Si and Mn show a linear effect on D_{0_eff} [3,23] while C shows an exponential influence, in the opposite direction to that reported by Mead and Birchenall [22]. Although it is noteworthy that the authors also assume activation energy varies with chemical composition. Table 3 collects the individual mean values D_{0_eff} of each alloy for the entire temperature range studied (900 to 1100°C).

Table 3. D_{0_eff} values for the all alloys studied.

	A000	A001	A010	A011	A100	A101	A110	A111
D_{0_eff} × 1·E-05	1.158	0.766	0.906	0.598	2.482	1.784	1.933	1.494

The idea that substitutional solutes, such as Si and Mn, can affect linearly the diffusivity behavior is further supported by Lidiard [25]. See Eqn. (4). Lidiard [25], for FCC materials and a dilute alloy, describes the self-diffusion of atoms in the presence substitutional solute as a linear function of the concentration of the solute element. Irmer et al. [26] point out that this relationship applies to substitutional solutes, although the authors have found equivalent relationships for interstitial solutes. Darken [27] established, for C diffusion in austenite, that the effect of Mn is negligible. Nevertheless Si markedly decreases the activity coefficient of C [27]. For a given C content, the presence of Si can retard the C mobility and this causes a decrease in the self-diffusion of Fe by affecting the lattice parameter of austenite [28-30]. It can be concluded that for a given C content, the addition of Si and Mn causes a decrease in pre-exponential factor and the largest decrease occurs for higher values of Si and Mn. Furthermore an increase of C content causes a marked increase in the pre-exponential factor.

The ability and goodness of the fitting is assessed using the correlation coefficient (Pearson's coefficient), the average absolute relative error AARE together with the root mean square deviation RMSD, as follows:

$$R = \frac{\sum_{i=1}^{N}\left(\sigma_{p\_ei} - \sigma_{p\_e\_m}\right)\left(\sigma_{p\_ci} - \sigma_{p\_e\_c}\right)}{\sqrt{\sum_{i=1}^{N}\left(\sigma_{p\_ei} - \sigma_{p\_e\_m}\right)^2 \sum_{i=1}^{N}\left(\sigma_{p\_ci} - \sigma_{p\_c\_m}\right)^2}} \quad (5)$$

$$AARE = \frac{1}{N}\sum_{i=1}^{N}\left|\frac{\sigma_{p\_ei} - \sigma_{p\_ci}}{\sigma_{p\_ei}}\right| \text{ x } 100 \quad (6)$$

$$RMSD = \sqrt{\frac{1}{N}\sum_{i=1}^{N}\left(\sigma_{p\_ei} - \sigma_{c\_ei}\right)^2} \quad (7)$$

Where σ_{p_ei} and σ_{p_ci} are the experimental peak stress and the calculated peak stress values (using Eqns. (1)-(4)) respectively. $\sigma_{p\_e\_m}$ and $\sigma_{p\_c\_m}$ are the mean values of the experimental and calculated values and N the number of data points respectively.

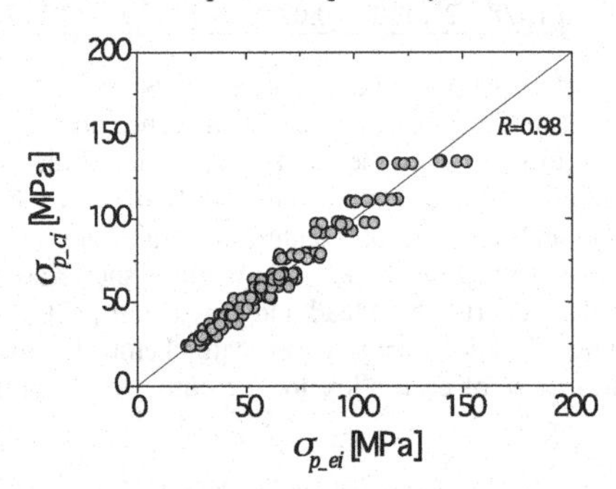

Fig. 4. Correlation between the experimental and predicted flow stress data.

As can be seen from Fig. 4, a good correlation between experimental and predicted data is obtained. Comparatively larger scattering in the data points above 140-150 MPa can be seen. This is probably due to adiabatic heating and friction for the experiments at the higher strain rates and the lower temperatures [31]. In spite of this scattering, R, the AARE and the RMSD are found to be 0.98, 6.2% and 5.5 MPa respectively for all the analyzed data. This fact reflects the good prediction capabilities of the proposed constitutive equation. The correlation coefficient is frequently used as a statistical indicator of the strength of linear relationship between the experimental and the calculated values. However, higher value of R may not necessarily indicate a better performance [32]. R, AARE and RMSD are computed through a term-by-term comparison of the relative error and therefore are unbiased statistics for measuring the predictability of a model [33]. The values obtained are in agreement with those reported in literature [34]. The model cover four orders of magnitude in strain rate for five values of temperature, always within the austenitic range, and, moreover, has not been assessed any correction for adiabatic heating.

CONCLUSIONS

The influence of C, Si and Mn on the pre-exponential factor D_{0_eff} of the self-diffusion coefficient for austenite is studied and an expression is derived (Eqn. (3)). This expression establishes a relation between $D_{0\_eff}^{1/n}$ and the chemical composition where carbon exerts the biggest influence, followed by silicon and manganese in decreasing order of importance.

ACKNOWLEDGMENTS

The authors are grateful to the Ministerio de Fomento of Spain for the support through the project N° 80032/A04. GVC is grateful for the research grant provided by the Comissionat per a Universitats i Recerca del Departament d´Innovació, Universitats i Empresa de Catalunya i del Fons Social Europeu. The authors wish to thank the support in the preparation of steels given by Dr. Sebastián Medina from CENIM, Madrid, Spain.

REFERENCES

1. U.R. Kattner and C.E. Campbell, *Mater. Sci. Technol.* **25(4)** (2009), 443.
2. H. Bakker et al. in "Diffusion in Solid Metals and Alloys" 1st Ed. vol. 26 Springer, (1990).
3. W.F. Gale and T.C. Totemeier, in "Smithell Metals Reference Book", 8th Ed Butterworth-Heinemann, (2004).
4. S-J. Lee, D.K. Matlock and C.J. van Tyne, *ISIJ International* **51(11)** (2011), 1903.
5. A. Ochsner, J. Gegner and G. Mishuris, *Met. Sci. Heat Treat.* **46(3-4)** (2004), 178.
6. J. Ågren, *J. Phys. Chem. Solids* **43(4)** (1982), 385.
7. J. Ågren, *J. Phys. Chem. Solids* **43(5)** (1982), 421.
8. J.O. Andersson and J. Ågren, *J. Appl. Phys.* **72(4)** (1992), 1350.
9. H.J. Frost and M.F. Ashby, in "Deformation-Mechanism Maps" 1st Ed. Pergamon Press, (1982).
10. T. Sakai and J.J. Jonas, *Acta Metall.* **32(2)** (1984), 189.
11. Y. Bergström, *Mat. Sci. Eng.* **5(4)** (1969), 193.
12. Y. Bergström and B. Aronsson, *Metall. Trans.* **3(7)** (1972), 1951.
13. U.F. Kocks, *J. Eng. Mat. Tech.* **98(1)** (1976), 76.
14. Y. Estrin and H. Mecking, *Acta Metall.* **32(1)** (1984), 57.

15. A.N. Kolmogorov, *Izvestija Akademii Nauk SSSR. Serija Matematiceskaja* **1** (1937), 355.
16. W.A. Johnson and R.F. Mehl, *Trans. Am. Inst. Min. Metall. Eng. AIME* **135** (1939), 416.
17. M. Avrami, *J. Chem. Phys.* **7(12)** (1939), 1103.
18. M. Avrami, *J. Chem. Phys.* **8(2)** (1940), 212.
19. J.M. Cabrera, A. Al Omar, J.J. Jonas and J.M. Prado, *Metall. Trans. A* **28A** (1997), 2233.
20. R.W. Cahn and P. Haasen, in "Physical Metallurgy", vol. I,II & III 4th ed., North Holland, (1996).
21. C. Zener, *J. Appl. Phys.* **22(4)** (1951), 372.
22. H.W. Mead and C.E. Birchenall, *Trans. Am. Inst. Min. Metall. Eng. AIME* **206(10)** (1956), 1336.
23. D. Treheux, L. Vincent and P. Guiraldenq, *Acta Metall.* **29(5)** (1981), 931.
24. F.S. Buffington, K. Hirano and M. Cohen, *Acta Metall.* **9(5)** (1961), 434.
25. A.B. Lidiard, *Philos. Mag.* **5(59)** (1960), 1171.
26. V. Irmer and M. Feller-Kniepmeier, *Philos. Mag.* **25(6)** (1972), 1345.
27. L.S. Darken, *Trans. Am. Inst. Min. Metall. Eng. AIME* **180** (1949), 430.
28. D.J. Dyson and B. Holmes, *J. Iron Steel Inst.* **208(5)** (1970), 469.
29. L. Cheng, A. Böttger, T.H. de Keijser and E.J. Mittemeijer, *Scripta Mater.* **24(3)** (1990), 509.
30. M. Onink et al., *Scripta Mater.* **29(8)** (1993), 1011.
31. R.L. Goetz and S.L. Semiatin, *J. Mater. Eng. Perform.* **10(6)** (2001), 710.
32. M.P. Phaniraj and A.K. Lahiri, *J. Mater. Process Technol.* **141(2)** (2003), 219.
33. S. Srinivasulu and A. Jain, *Appl. Soft Comput.* **6(3)** (2006), 295.
34. S. Mandal, V. Rakesh, P.V. Sivaprasad, S. Venugopal and K.V. Kasiviswanathan, *Mat. Sci. Eng. A* **500(1-2)** (2009), 114.

Mater. Res. Soc. Symp. Proc. Vol. 1485 © 2013 Materials Research Society
DOI: 10.1557/opl.2013.207

Recent Advances on Bulk Tantalum Carbide Produced by Solvothermal Synthesis and Spark Plasma Sintering

Braeden M. Clark,[1] James P. Kelly,[1] and Olivia A. Graeve[1,*]
Kazuo Inamori School of Engineering, Alfred University
Alfred, NY 14802, U.S.A.

ABSTRACT

The sintering of tantalum carbide nanopowders by spark plasma sintering (SPS) is investigated. The washing procedure for the powders is modified from previous work to eliminate excess lithium in the powders that is left over from the synthesis process. The sintering behavior of the nanopowders is investigated by X-ray diffraction and scanning electron microscopy by studying specimens that were sintered to different temperatures. To improve the homogeneity of the microstructure of the specimens, milling procedures were implemented. Vaporization during sintering is observed, and the usefulness of carbon additions and systematic decreases in temperature to curb this behavior was explored. Future experiments to achieve full density and to maintain a nanostructure of the specimens include sintering with higher pressures, lower temperatures, and longer dwell times. Additives for maintaining a nanostructure and developing suitable high-temperature properties are also discussed.

INTRODUCTION

The following work describes the process of sintering tantalum carbide in an attempt to obtain dense and nanostructured specimens. Tantalum carbide is one of the most refractory ceramics known with a melting temperature of around 3800°C. Therefore, this material is potentially useful for space reentry vehicles and high mach aircrafts, where extreme temperatures are reached. Other properties, including hardness, fracture toughness, elastic moduli, oxidation behavior, electronic structure, and optical properties, have also been investigated for this material.[1-4]

The preparation of bulk nanoceramics can result in improved material properties. Maintaining a nanostructure has been shown to modify electrical properties and mechanical properties, for instance.[5-7] Properties that did not exist in a material with grain sizes in the micrometer range can potentially exist when the grain size is reduced to the nanometer scale.[8] Spark plasma sintering is a useful technique to consolidate powder because of the simultaneous pressure and current applied to create fast heating rates and sintering cycles, which can substantially limit the amount of grain growth of a sintered specimen. This technique is investigated for sintering tantalum carbide nanopowders synthesized by a solvothermal synthesis method.

EXPERIMENTAL PROCEDURE

Previous sintering work revealed that the scaled-up synthesis of tantalum carbide powders, which were subsequently sintered, still contained lithium left over from the synthesis

[*] Author to whom correspondence should be addressed: Email: graeve@alfred.edu, Tel: (607) 871-2749, Fax: (607) 871-2354, URL: http://people.alfred.edu/~graeve/

process.[9] Tantalum carbide powders were synthesized using a solvothermal technique previously established to produce a 50 g theoretical yield of powder.[9-10] The synthesis method is rapid and could easily be scaled to produce large quantities of powder in a continuous type of process. After synthesis, the powders were oxidized by heat treating in air at 900°C for 1 hour and then characterized by X-ray diffraction (XRD), which revealed a lithium tantalate phase due to a residual lithium phase left over from the initial reaction.

The lithium is in the form of lithium hydroxide formed by the reaction of lithium with water when freeing the reaction products after the reaction cools. The synthesized powders were divided into two sets of powders for washing experiments to explore the removal of the excess lithium hydroxide. Lithium hydroxide is water soluble. Additions of nitric acid can form lithium nitrate from the lithium hydroxide, which has a higher solubility limit in water than lithium hydroxide and is a method that has been used to eliminate lithium hydroxide, magnesium hydroxide, and calcium hydroxide from powders made by a similar process.[11]

One set of powders was washed with water and the other set was washed with a dilute nitric acid solution. The nitric acid solution was made by mixing a volumetric ratio of 1:19 acid to water, using a 15.8 N nitric acid stock solution. Both sets of powders were washed in 500 mL of solution. One complete washing cycle consisted of 30 minutes of magnetic stirring, 30 minutes of ultrasonication, 15 minutes of magnetic stirring, and centrifuging, until the powders were separated and the supernatant could be poured off. A small portion of powder was oxidized after each wash, in the same manner as outlined previously, and XRD was performed to confirm whether or not the lithium tantalate phase, indicating residual lithium hydroxide in the powders, was present or not.

After lithium hydroxide-free powders were obtained, spark plasma sintering experiments were performed. One sample was sintered at a temperature of 2050°C at a heating rate of 100°C/min and a 50 MPa uniaxial pressure. All subsequent sintering trials used the same parameters, but with variations in sintering temperature. The resulting displacement curve showed 5 distinct displacement peaks at different temperatures. Sintering of samples to maximum temperatures of 800, 1100, 1500, and 1950°C for 30 seconds was carried out. XRD and scanning electron microscopy (SEM) was used to study the sintering behavior for each of these temperatures.

Due to microstructure irregularities that were observed after sintering, a milling processing step was implemented to homogenize the powders prior to sintering. The tantalum carbide powders were milled for two hours prior to being sintered to 2050°C and the microstructure was characterized and compared to the microstructure when sintering the powders that were not milled. In addition, one sample was sintered to 2200°C and held for ten minutes in an attempt to complete densification.

Carbon additions were made to the powders and sintered at 2200°C for 10 minutes to compare to sintering behavior of the as-synthesized and processed powders. Using the best results from the carbon addition experiments, the heating rate was investigated. Some common features were observed, including apparent vaporization and so reduction in temperature with a 10 minute hold time was also investigated.

RESULTS AND DISCUSSION

Relevant XRD patterns associated with the washing procedure are shown in Figure 1. It is clear that the as-synthesized powders contain a large proportion of lithium hydroxide from the lithium tantalate observed in the XRD pattern. Tantalum carbide powders washed with a dilute nitric acid solution did not contain lithium hydroxide (or residual lithium nitrate) after two washing cycles. When these powders were oxidized, a tantalum oxide phase without a trace of the previously observed lithium tantalate phase was observed in the XRD pattern.

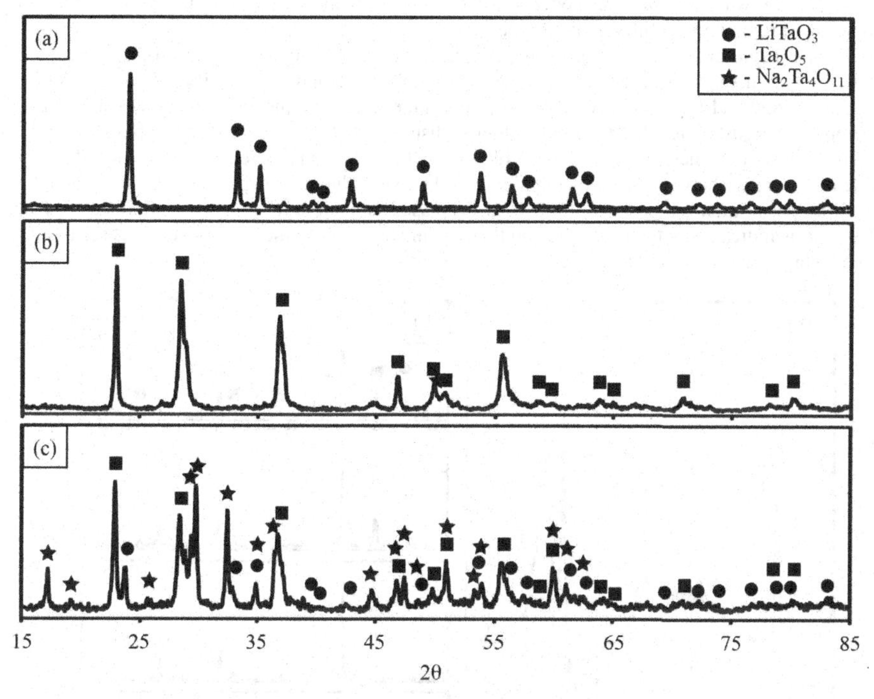

Figure 1. Powder XRD patterns for oxidized (a) as-synthesized powders, (b) powders after two nitric acid washing cycles, and (c) powders after nine water washes that show patterns for phase-pure lithium tantalate, phase-pure tantalum oxide, and a mixture of phases, respectively.

In comparison, powders washed only with water still had lithium hydroxide even after nine washing cycles. After oxidation, the lithium tantalate phase was still observed in the XRD pattern, although the peaks are of lower intensity and the tantalum oxide phase is now also observed. A third phase was found and is indicated by additional peaks compared to the as-synthesized and nitric acid washed powders. The third phase is most likely an intermediate phase for the reaction of lithium hydroxide and tantalum carbide to form the lithium tantalate phase, deficient in lithium. This phase was strongly matched to natrotantite ($Na_2Ta_4O_{11}$), but since there is no sodium source the phase is likely a lithium-containing phase of similar structure although no such phase was available in the database. If we assume that the chemical formula for natrotantite is a guide, then the compound would have stoichiometry of $Li_2Ta_4O_{11}$, which is four $LiTaO_3$ units deficient in one Li_2O unit, agreeing with the assessment of an intermediate phase that is deficient in lithium.

Lithium hydroxide is less soluble in water than lithium nitrate and can explain the improved washing efficiency when using nitric acid. The observations agree with our hypothesis outlined in the experimental procedure and our previous work concerning the effect of nitric acid.[11] Water washing was abandoned after the ninth washing cycle due to the decreased efficiency of removing the lithium hydroxide.

The displacement rate curve for the sample sintered at 2050°C is displayed in Figure 2. There are five distinct displacement rate peaks of the curve that indicate densification: near 750, 925, 1300, 1650 and 1750 °C. Figure 2 also displays the XRD patterns for four samples sintered to temperatures just above those corresponding to displacement rate peaks. The sample sintered at 800°C shows the emergence of an oxide phase. At 1100°C, the relative peak intensities of the oxide and carbide phases indicate that the level of the oxide phase is increased. The presence of the oxide phase after sintering to 1500°C is decreased and the oxide phase is eliminated by 1950°C as is indicated by the reduction and then elimination of the oxide peak intensities relative to the tantalum carbide peak intensities.

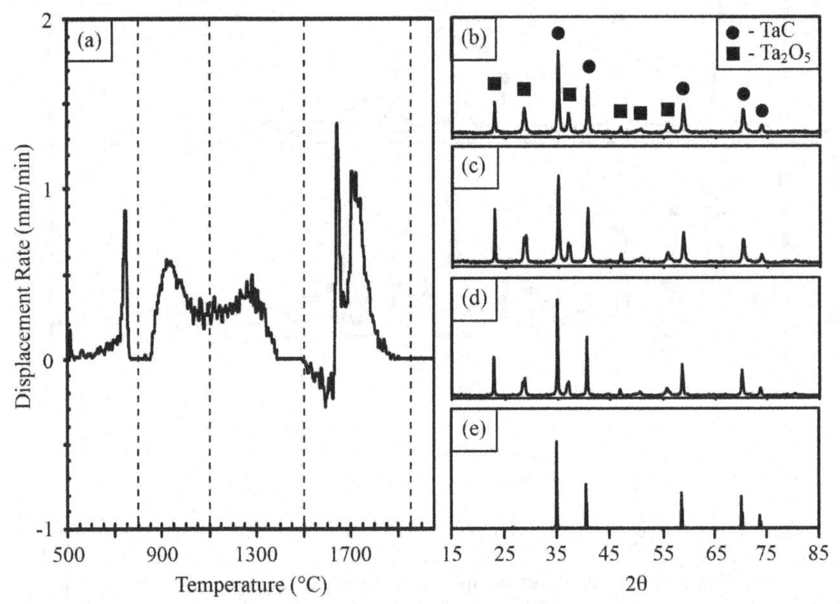

Figure 2. (a) Displacement rate curve for a sample sintered at 2050°C. Dotted lines in this curve are cut-off temperatures, held for 30 s, for the sintering of secondary specimens used to generate the X-ray diffraction patterns given in (b)-(e). The X-ray diffraction patterns are shown for the specimens sintered to temperatures of (b) 800°C, (c) 1100°C, (d) 1500°C, and (e) 1950°C.

Previous work indicated relatively high oxygen content (~6.1%) in the tantalum carbide powders synthesized by the solvothermal method [9]. This can explain the presence of the oxide phase at lower sintering temperatures. However, it is unclear whether the oxygen is a result of strongly bound surface molecules, an oxide scale on the surfaces, or even bulk lattice oxygen.

Despite the presence of an oxide phase at lower temperatures, it is eliminated at higher temperatures because excess carbon is introduced in the initial reaction. From the thermogravimetric analysis (TGA) data in our previous work, the weight loss during oxidation is ~2.4%.[9] A free carbon content of 2.4% could not fully react to eliminate 6.1% of oxygen via the evolution of carbon monoxide gas that might be expected from carbothermal reduction. However, the carbon content could be higher because the actual amount could be masked by the concurrent oxidation of the powders when performing the TGA measurements. Furthermore, tantalum carbide is capable of incorporating high levels of sub stoichiometry.[12] It is possible that lattice carbon is an additional carbon source that can react with oxygen to reduce the oxide phase as well.

SEM images of fracture surfaces for the specimens sintered to 800°C and 1500°C are shown in Figure 3. At 800°C, agglomerate structures having sub-micrometer to a few micrometer sized features are made up of very small nanostructured grains. A low density phase (shown as a dark and faint phase) is also observed in this specimen. This low density phase is not seen at 1100°C or above. The low density phase may be residual lithium hydroxide. Lithium hydroxide decomposes just above 900°C and would explain why it is not observed in the other specimens. In small concentrations, the lithium hydroxide may not be detected as lithium tantalate after oxidation because the concentration is below the detection limits of XRD. The presence of oxide in the sample at this temperature suggests that the sintering displacement could be related to the formation of the oxide phase. It is also near the temperature for which free carbon and readily available oxygen (not lattice oxygen) can form carbon monoxide gas which may be responsible for the observed displacement.

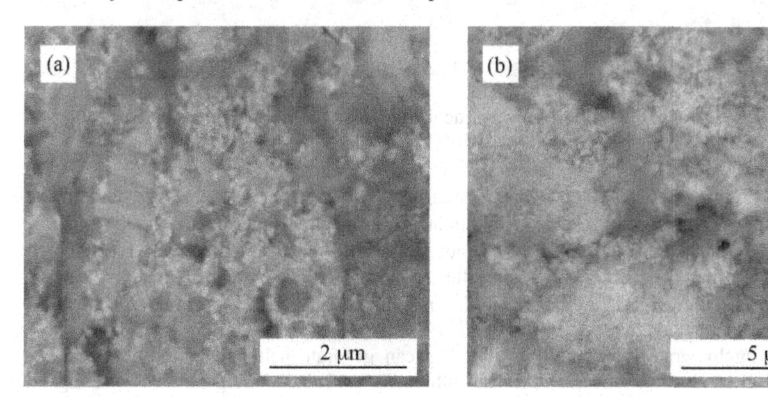

Figure 3. Back-scattered electron images of fracture surfaces for the specimens sintered at (a) 800°C and (b) 1500°C for 30 seconds.

Signs of sintering begin to show by 1100°C, but the sintering behavior is non-uniform. The behavior is enhanced for sintering up to 1500°C, which is depicted in Figure 3. Regions of slightly lower average atomic number (darker phase in the image) are sintered into large (several micrometers) and irregular grains. Amongst these large grains are submicron-sized particles of a brighter phase, presumably the tantalum carbide grains of higher average atomic number. The second and third displacements are likely due to sintering of the oxide phase in the compact to form the large irregular grains.

Figure 4 displays SEM images of the specimens sintered at 1950°C and 2050°C. After sintering to 1950°C, the large scale grains presumed to be the oxide phase in Figure 3 are no longer observed in the specimens. The elimination of the oxide phase observed in the XRD analysis, shown in Figure 2, agrees with this assessment. These results suggest that the negative displacement just after 1500°C, and the two positive displacement peaks at 1650°C and 1750°C correspond to the carbothermal reduction of the oxide phase and sintering of the tantalum carbide. The initial negative displacement could be expansion due to the formation of carbon monoxide, followed by a positive displacement corresponding to the sintering of the tantalum carbide grains.

Figure 4. Back-scattered electron images of fracture surfaces for the specimens sintered at (a) 1950°C and (b) 2050°C for 30 seconds.

When sintering at 1950°C or 2050°C, significant grain growth is observed and the nanostructure is lost. The grain size is approximately 1 μm after sintering to 1950°C. By comparing the two images in Figure 4, it is obvious that higher sintering temperature results in increased grain growth. The average grain size after sintering at 2050°C is approximately 2 μm. The increase in sintering temperature does not eliminate the porosity in the specimen to achieve full density.

A large agglomerate distribution in powders can generate heterogeneous green body microstructures after forming processes that can ultimately lead to non-homogenous sintering. This is demonstrated in Figure 5. This figure shows an agglomerate of approximate dimensions of 20 μm wide by 40 μm long. In this case, the agglomerate sintered slightly faster than the surrounding bulk material due to increased particle-particle contact within the agglomerate compared to the inter agglomerate particle-particle contacts. The faster sintering causes the agglomerate to shrink away from the surrounding agglomerates and generates a high density of pores surrounding the agglomerate. For much larger agglomerates, this effect is amplified. The large agglomerate of the same sample in Figure 5 that is approximately 50 μm wide by 100 μm long shrinks away from the surrounding material matrix to such an extent that the pore network surrounding the agglomerate has evolved to a microstructural crack surrounding the agglomerate.

Forming a narrow agglomerate size distribution of powder agglomerates is an important feature of a forming process for avoiding such defects as that demonstrated in Figure 5. One of the most common techniques for controlling agglomerate size and distribution is by spray drying, but this technique requires some knowledge of how the powders behave in suspension and manipulating the spray drying characteristics using particle dispersion techniques and/or binder and plasticizer additives. A slightly more rudimentary, but effective approach, is dry powder cascade milling. Excessively large agglomerates will be broken down and excessively small agglomerates will grow during milling, forming a narrower agglomerate size distribution.

Figure 5. Image (a) shows a high pore density network surrounding an agglomerate due to differential shrinkage during sintering. Larger agglomerates in the microstructure, such as shown in (b), show that if the agglomerate shrinkage away from the surrounding matrix is large enough, then the pore channel can evolve into a microstructural crack. Image (c) is a closer look at the crack from image (b).

If the powders are milled prior to sintering, then the detrimental cracking of the sample is avoided. An example is shown in Figure 6. Although pore clusters are not avoided altogether, the microstructural heterogeneity is less notable and the heterogeneity appears to at least be consistent throughout the entire specimen microstructure. Further optimization of the agglomerate structure and size distribution may assist in achieving full density of the samples. A homogenous green body microstructure will sinter and shrink isotropically. Isotropic shrinkage conditions will not cause two areas of the microstructure to separate from each other like anisotropic shrinkage does and pores will close uniformly instead of pore closure within the agglomerates and pore growth amongst the agglomerates.

One sample was sintered to 2200°C and was held at that temperature for 10 minutes in an attempt to see if increased temperature and sintering time could eliminate porosity in the achieved microstructure. Figure 7 displays SEM images of this sample. Elimination of the porosity was not achieved. Grain growth also continued throughout the sintering time and grains approaching a size of approximately 10 μm were observed.

Notable observations when sintering at 2200°C are shown in Figure 8. Many of the grains have a complex faceted morphology. On the center of some of the larger flat faces appear to be spherical cavities that can be observed in both secondary electron and back-scattered electron imaging modes. Another common feature that can be observed is grain texturing, which can be as simple as a set of straight striations across the face of a grain or more complex sets of striations with curvature that provide complex grain morphologies.

The features outlined in Figure 8 can be the result of vaporization of material. Vapor transport is detrimental to densification. Oxygen content in the powders can lead to vaporization during the sintering process and hinder densification. Carbon additions of 3, 5, and 6 weight percent and sintering of these powders to 2200°C for 10 minutes for purposes of eliminating the oxygen were made. Based on an estimated oxygen content of ~6.1% and a free carbon content of 2.6%, we estimate that 2.6% carbon addition can remove the oxygen via CO formation. An addition of 3% should be enough to remove oxygen based on this computation. Additions of 5 or 6% represent just under and just over the 5.2% addition of carbon that is estimated for removal of the oxygen and replacing the oxygen with carbon.

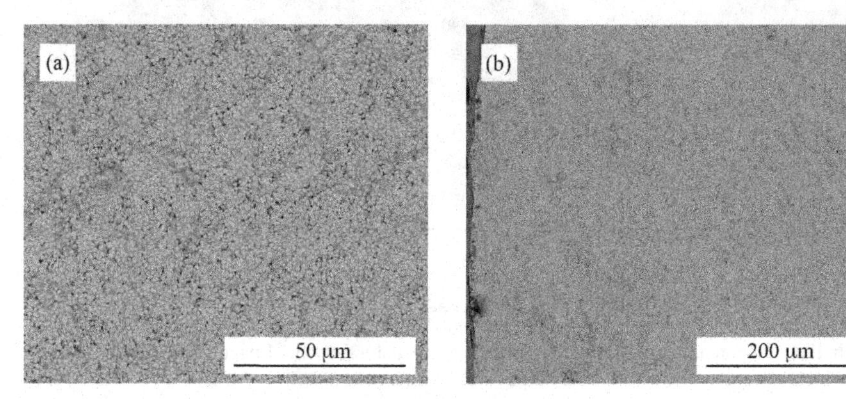

Figure 6. Image (a) shows that sintered powders that were milled still show some heterogeneity in porosity, but on a large scale such as that shown in image (b), the overall microstructure is homogeneous and free of cracks.

The microstructures of samples with and without carbon additions are shown in Figure 9. Without carbon additions, observations remain similar to those described before. A 3% carbon addition reduced the amount of spherical cavities in the grains, although they were not completely eliminated. Furthermore, the grains have improved morphology, becoming more equiaxed. Although some carbon additions were beneficial, further additions of 5% or 6% carbon was detrimental. The grains continue to grow, grain morphologies are more irregular than without carbon additions, and there are even greater levels of the spherical cavities in the grains. With the excess addition of carbon, the larger grains are also accompanied by larger porosity. For the 6% sample, microstructure heterogeneity was also observed where residual carbon collects and remains in the microstructure. For further experiments 3% carbon additions were used.

Faster heating rates were also investigated for eliminating the porosity of the specimens. The microstructures obtained by using various heating rates are shown in Figure 10. It was determined that the heating rate is an insignificant sintering variable when considering the densification behavior of these samples. Another method of reducing vaporization is to lower the sintering temperature (the phenomenon observed in Figure 8 is not present in Figure 4).

Systematic decreases in temperature, maintaining a 10 minute dwell time, was performed to attempt to eliminate porosity.

Figure 7. Back-scattered electron image of a fracture surface of a specimen sintered at 2200°C taken at initial magnifications of (a) 2000x and (b) 200x. The lower magnifications shows that the microstructure remains homogeneous throughout.

Figure 8. Back-scattered electron images showing common features observed when sintering to 2200°C include (a) spherical cavities on flat grain faces and (b) striations that mark the grain faces.

Figure 11 shows the microstructures for samples sintered at systematically lower temperatures for 10 minutes. The grain sizes and pore sizes seem to be reduced by lowering the sintering temperature, but the porosity and vaporization are not completely eliminated by dropping the sintering temperature to 2050°C. A further reduction in sintering temperature could continue to improve the densification of the powders. By comparing Figure 11c and Figure 4b, both sintered at 2050°C, it becomes apparent that the combination of longer sintering time, carbon addition, and increased sintering of the sample has lead to significant improvement in the development of a higher density microstructure.

It is worth noting that sintering time is a variable that has not been studied yet and is a focus of current ongoing work in our laboratory. A uniaxial pressure of 50 MPa was used throughout this study and higher pressures may also promote densification. When full density of the specimens is achieved, then research can be focused on maintaining a nanostructure. The process of achieving full density might even lead to a nanostructure naturally. The average crystallite size of the starting powder is approximately 25 nm and the average particle size near 100 nm. Preventing grain growth will produce microstructures with grain sizes of similar magnitude.

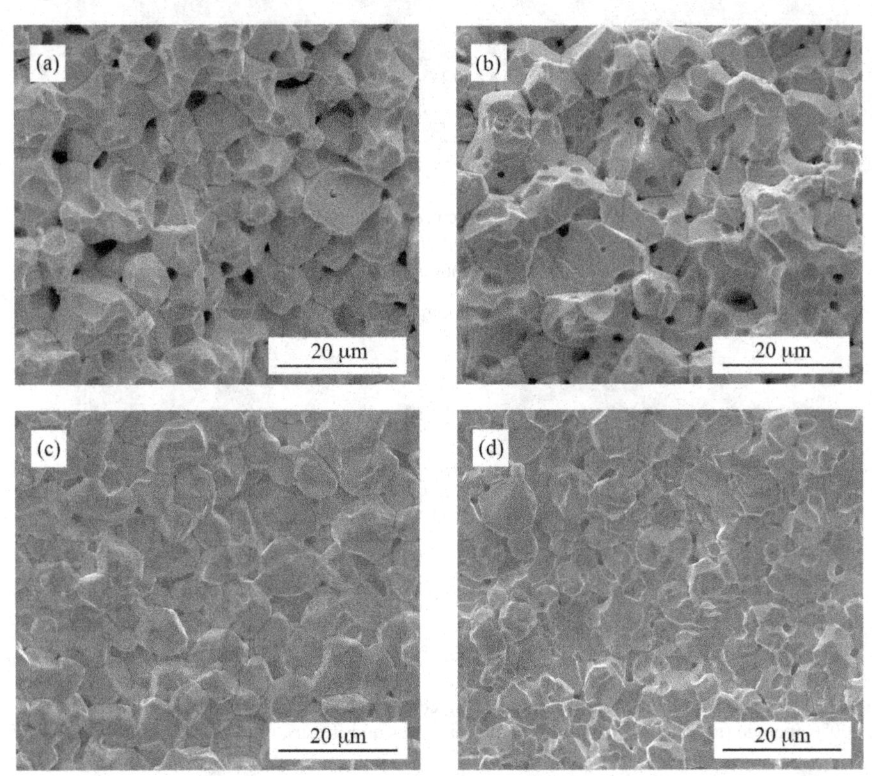

Figure 9. Secondary electron images of specimens sintered at 2200°C (a) without carbon addition, (b) with 3% carbon addition, (c) with 5% carbon addition, and (d) with 6% carbon addition.

Tantalum carbide is limited by its oxidation resistance. The design of multi-phase materials that upon oxidation will form a stable (compatible thermal expansion to prevent cracking and spalling) and high-melting temperature oxide scale would have a high-temperature advantage over single phase tantalum carbide. The design of an oxide scale that becomes semi-viscous at an application temperature may also be useful, so that the scale becomes self-healing when damaged. If such an additive is also insoluble in tantalum carbide and vice versa, then the additive can act as a grain growth inhibitor. The use of grain growth inhibitors may be useful in controlling the grain growth of the samples to keep them nanostructured.

Figure 10. Secondary electron images of specimens sintered at 2200°C, but using heating rates of (a) 200°C/min, (b) 300°C/min, and (c) 400°C/min.

Figure 11. Secondary electron images of specimens sintered with 3% carbon addition and held for 10 minutes at temperatures of (a) 2150°C, (b) 2100°C, and (c) 2050°C using a heating rate of 400°C/min.

CONCLUSIONS

The spark plasma sintering of tantalum carbide powders produced by solvothermal synthesis was investigated after optimizing the washing procedure to remove lithium hydroxide from the powders. The sintering characteristics were examined under various conditions. Dry powder milling, sintering temperature, carbon additions, and sintering hold time are important variables for sintering and were used to make significant advances in attempts to achieve full density compacts from the powders. Further optimization of the washing procedures, to reduce the oxygen content, may be useful, but was found to not be necessary for obtaining phase pure tantalum carbide after sintering.

Current work is geared towards obtaining full density specimens. Higher pressures, lower temperatures, and longer sintering times are currently being investigated. If full density can be achieved, then research and development can be shifted towards maintaining a nanostructure.

ACKNOWLEDGEMENTS

This work was financially supported by a grant from the National Science Foundation under grant numbers CMMI 0645225 and CMMI 0913373.

REFERENCES

[1] A. Dashora and B.L. Ahuja, Electronic Structure, Compton Profiles and Optical Properties of TaC and TaN, *Radiat. Phys. Chem.*, **79**, 1103-10 (2010).

[2]. M. Desmaison-Brut, N. Alexandre, and J. Desmaison, Comparison of the Oxidation Behaviour of Two Dense Hot Isostatically Pressed Tantalum Carbide (TaC and Ta_2C) Materials, *J. Eur. Ceram Soc.*, **17**, 1325-34 (1997).

[3]. I.-J. Shon, D.-M. Lee, J.-M. Doh, J.-K. Yoon, and I.-Y. Ko, Consolidation and Mechanical Properties of Nanostructured $MoSi_2$–SiC–Si_3N_4 from Mechanically Activated Powder by High Frequency Induction Heated Sintering, *Mater. Sci. Eng., A*, **528**, 1212-5 (2011).

[4]. L. Silvestroni, A. Bellosi, C. Melandri, D. Sciti, J.X. Liu, and G.J. Zhang, Microstructure and Properties of HfC and TaC-Based Ceramics Obtained by Ultrafine Powder, *J. Eur. Ceram. Soc.*, **32**, 619-27 (2011).

[5]. H. Hahn and R.S. Averback, High Temperature Mechanical Properties of Nanostructured Ceramics, *Nanostruct. Mater.*, **1**, 95-100 (1992).

[6]. T. Hungría, H. Amorín, M. Algueró, and A. Castro, Nanostructured Ceramics of $BiSCO_3$–$PbTiO_3$ with Tailored Grain Size by Spark Plasma Sintering, *Scr. Mater.*, **64**, 97-100 (2011).

[7]. D.G. Lamas, M.F. Bianchetti, M.D. Cabezas, and N.E.W. de Reca, Nanostructured Ceramic Materials: Applications in Gas Sensors and Solid-Oxide Fuel Cells, *J. Alloys Compd.*, **495**, 548-51 (2010).

[8]. P. Bowen and C. Carry, From Powders to Sintered Pieces: Forming, Transformations and Sintering of Nanostructured Ceramic Oxides, *Powder Technol.*, **128**, 248-55 (2002).

[9]. B.M. Clark, J.P. Kelly, and O.A. Graeve, Exploring the Synthesis Parameters and Spark Plasma Sintering of Tantalum Carbide Powders Prepared by Solvothermal Synthesis, *Mater. Res. Soc. Symp. Proc.*, **1373**, 7-17 (2012).

[10]. J.P. Kelly, R. Kanakala, and O.A. Graeve, A Solvothermal Approach for the Preparation of Nanostructured Carbide and Boride Ultra-high Temperature Ceramics, *J. Am. Ceram. Soc.*, **93**, 3035-8 (2010).

[11]. J.P. Kelly and O.A. Graeve, Statistical Experimental Design Approach for the Solvothermal Synthesis of Nanostructured Tantalum Carbide Powders, *J. Am. Ceram. Soc.*, **94,** 1706-15 (2011).

[12]. A.L. Bowman, The Variation of Lattice Parameter with Carbon Content of Tantalum Carbide, *J. Phys. Chem.*, **65**, 1596-8 (1961).

Mater. Res. Soc. Symp. Proc. Vol. 1485 © 2013 Materials Research Society
DOI: 10.1557/opl.2013.208

Removal of Lead Ions in an Aqueous Medium with Activated Carbon Fibers

Teresa Ramírez-Rodríguez and Fray de Landa Castillo-Alvarado
Escuela Superior de Física y Matemáticas, Instituto Politécnico Nacional, 07738 Ave. IPN, D.F, México.

ABSTRACT

Poly(acrylonitrile) fibers are used in the manufacture of activated carbon fibers, which are activated with phosphate groups for the removal of lead ions in aqueous solutions. Removal of lead ions is performed in a water bath at 30°C. Trough isotherm models of Langmuir and Freundlich analyzed the aqueous solution. Kinetic analysis is performed using the model pseudo-first and pseudo-second order. The result show that adsorption equilibrium is adjusted to the Freundlich model and the kinetic model of pseudo-second order led to the best fit correlation.

INTRODUCTION

Different treatments for activated carbon adsorption are commonly used due to its relatively high performance and safety in the environment [1]. Activated carbon is manufactured in order to provide a high degree of porosity and high specific surface. In this paper have been used poly(acrylonitrile) fiber precursor material to produce activated carbon fibers (ACF). The aim of this study is to investigate the kinetic of adsorption and removal capacity in aqueous solutions of lead ions by ACF derived from acrylic fibers.

THEORY

Relationship Aqueous solution:mass of fiber is indicated as: 1:1, 1.5:1, 2:1, 3:1, 1:2, 1:3, and 1:1.5., [2mL:0.1g]. Removal of lead ions has been carried out in an aqueous bath at a temperature of 30°C. Samples are prepared by filtration and concentration of lead ions (II) is analyzed by an atomic absorption spectrophotometer. Each experiment is doubled under the same conditions. The following models are used for isothermal and kinetic analysis.

Langmuir isotherm

In 1916 Langmuir [2] developed a simple model to predict the adsorption of a gas on a surface as a function of fluid pressure. However, this model is applicable to solid-liquid interface as the solid-gas interface [3]. Liu [4] proposed a dynamic schema that represents the adsorption:

$$A + B \leftrightarrow AB \qquad (1)$$

Where A represents the adsorbate, B represents the solid adsorbent and AB represent the adsorbent-adsorbate complex. For equation (1) the equilibrium constant can be expressed as:

$$K = \frac{[AB]}{[A][B]} \qquad (2)$$

According to the definition of equilibrium constant [AB], [A] [B] are the molar concentrations of each component. The number of moles of adsorbate bound per mole of adsorbent (q) can be written as:

$$q = \frac{[AB]}{[B]+[AB]} \quad (3)$$

Substituting equation (2) in equation (3), we have:

$$q = \frac{K[A]}{1+K[A]} \quad (4)$$

Being 1/K a dissociation constant, on the other hand, if the adsorbent has n identical binding sites, the adsorption isotherm can be considered as the sum of each of the binding sites:

$$q = n\frac{K[A]}{1+K[A]} \quad (5)$$

Equation (6) shows that the theoretical maximum adsorption capacity is n mol mol^{-1}. Multiplying both sides of equation (5) by the ratio of the molar weight of adsorbate (M_A) and the molar weight of adsorbent (M_B) gives:

$$\frac{M_A}{M_B} q = n\frac{M_A}{M_B}\frac{K[A]}{1+K[A]} \quad (6)$$

The term $\left(M_A/M_B\right)q$ is considered Q and the term $n\left(M_A/M_B\right)$ is considered Q_m, so equation (6) is arranged as:

$$Q = \frac{Q_m K[A]}{1+K[A]} \quad (7)$$

Where $Q\left(g\,A\,g^{-1}\,B\right)$ is the adsorption capacity, $Q_m\left(g\,A\,g^{-1}\,B\right)$ is the maximum adsorption capacity and K is the equilibrium constant of adsorption $\left(L\,mol^{-1}\right)$. The equation (8) is known as the Langmuir isotherm. It should be emphasized that [A] in equation (7) is the molar concentration of adsorbent A in the balance. In practice it is often to express true concentrations of adsorbate in equilibrium with different units to molar concentration, i.e., $mg\,L^{-1}$ o $\mu g\,L^{-1}$ among others. In these cases the nomenclature of the isotherm and the units of Langmuir adsorption constant change as show in equation (8).

$$q_e = \frac{q_m b C_e}{1+bC_e} \quad (8)$$

Where q_e is the amount of solute adsorbent per unit weight of adsorbent at equilibrium $(mg\,g^{-1})$, q_e is the equilibrium concentration of solute in the volume of the solution $(mg\,L^{-1})$, q_m is the maximum adsorption of Pb (II) $(mg\,g^{-1})$, and b is the constant related to the free energy of adsorption $(L\,mg^{-1})$.

Freundlich isotherm

The expression of the Freundlich isotherm is an exponential equation that assumes that when the adsorbate concentration increases so does the concentration of adsorbate in the adsorbent surface [5, 6].

$$q_e = k_F\, C_e^{1/n} \qquad (9)$$

Where k_F a constant indicative of the relative adsorption capacity of is adsorbent $\left(mg^{1-(1/n)}\, L^{1/n}\, g^{-1}\right)$ and n is a constant indicative of adsorption intensity and $1/n$ is a measure of the heterogeneity of the adsorptive surface, this constant take values between 0 and 1. There is a greater surface heterogeneity of the value of $1/n$ approaches to 0 [7], if $1/n < 1$ the adsorption is favorable and if $1/n > 1$ is unfavorable. Theoretically Freundlich expression suggests that it maybe occur an infinite amount of adsorption [6]. To determine the parameters of k_F and $1/n$ using the linear form of Freundlich equation and where the constants are determined using the slope and intercept obtained by linear regression.

$$\ln(q_e) = \ln(k_F) + 1/n \ln C_e \qquad (10)$$

Kinetic model of pseudo-first order

The rate equation of adsorption of pseudo-first order Lagergren is originally designed in 1898 by adsorption systems for liquid-solid. Lagergren studied the adsorption of oxalic acid and malonic into charcoal. The adsorption rate expression resulting from these studies is usually expressed as follows [8].

$$\frac{dq_t}{dt} = k_1 (q_e - q_t) \qquad (11)$$

Where q_e and q_t are the equilibrium adsorption capacity and a t, respectively $(mg\,g^{-1})$, k_1 is the constant rate of adsorption pseudo-first order $(L\,\min^{-1})$. If equation (11) is integrated with boundary conditions by $t = 0$, $q_0 = 0$ and $t = t_1, q_t = q_1$. It has obtained the integrated reaction rate of adsorption pseudo-first order, which becomes:

$$\ln\left(\frac{q_e}{q_e - q_t}\right) = k\,t \qquad (12)$$

$$q_t = q_e\left(1 - e^{-kt}\right) \quad (13)$$

$$\ln\left(\frac{q_e}{q_e - q_t}\right) = kt \quad (14)$$

From equation (14) has obtained the linear form:

$$\log(q_e - q_t) = \log(q_e) - \frac{k}{2.303}t \quad (15)$$

From equation (15) kinetic constant are found k and q_e

Kinetic model of pseudo-second order

In 1995, Ho presented the adsorption rate expression of pseudo-second order, which shows how the rate of adsorption depends on the ability of equilibrium adsorption and the concentration of the adsorbate [9]. In an attempt to represent the equation of adsorption divalent metals in a material during agitation, it is assumed that the process can be second order and considering that the chemical adsorption involving valence forces through the exchange of electrons between the material and divalent metal ions involve covalent forces, the second-order reaction may depend on the amount of divalent metal ions on the surface of the material and the amount of divalent metal ions adsorbed at equilibrium. The rate expression for adsorption is described:

$$\frac{d(P)_t}{dt} = k\left[(P)_0 - (P)_t\right]^2 \quad (16)$$

$$\frac{d(HP)_t}{dt} = k\left[(HP)_0 - (HP)_t\right]^2$$
(17)

Where $(P)_t$ and $(HP)_t$ are the number of active sites occupied in the material at time t, and $(P)_0$ and $(HP)_0$ are the number of sites available on the balance of the material.

The driving force is proportional to the available fraction of active sites. The adsorption rate equations can be rewritten as follows [10].

$$\frac{dq_t}{dt} = k_2(q_e - q_t)^2 \quad (18)$$

Where k_2, is the rate constant of adsorption $\left(g\ mg^{-1} \min^{-1}\right)$, q_e is the amount of adsorbed divalent metal ions at equilibrium $\left(mg\ g^{-1}\right)$ and q_t is the amount of divalent metal ions on the surface adsorbent anytime, t, $mg\ g^{-1}$. Separating in equation (18):

$$\frac{dq_t}{\left(q_e - q_t\right)^2} = k\,dt \tag{19}$$

Integrating over the boundary conditions: $t = 0$, $q_0 = 0$ and $t = t_1$, $q_t = q_1$

$$q_t = \frac{q_e^2 k t}{1 + q_e k t} \tag{20}$$

Equation (20) is the integrated rate law for a reaction of adsorption pseudo-second order. The equation in its linear form:

$$q_t = \frac{t}{\dfrac{1}{kq_e^2} + \dfrac{t}{q_e}} \tag{21}$$

$$\frac{1}{q_t} = \frac{1}{kq_e^2} + \frac{1}{q_e} t \tag{22}$$

$$h = kq_e^2 \tag{23}$$

Where h is the initial adsorption rate $\left(mg\ g^{-1} \min^{-1}\right)$ as q_t/t approaches to 0, and equation (21) can be arranged for:

$$q_t = \frac{1}{\dfrac{1}{h} + \dfrac{1}{q_e}} \tag{24}$$

$$\frac{t}{q_t} = \frac{1}{h} + \frac{1}{q_e} t \tag{25}$$

The constants for the model of pseudo-second can be determined experimentally by plotting t/q_t vs t.

DISCUSSION

After subjecting the ACF by atomic absorption spectrometry, resulting that the maximum efficiency in the uptake of ions lead is determined in a contact time of 120 minutes, the percentage of removal becomes an average of 99.90 percent.

Adsorption isotherm of Langmuir

Following graph is obtained using a linear form of Langmuir adsorption model. Figure 1 shows that the samples in the following order with the best performance of the removal capacity of lead ions with respect to the equilibrium solute content in the aqueous solution; 1:2, 1.5:1, 1:1, 2:1.

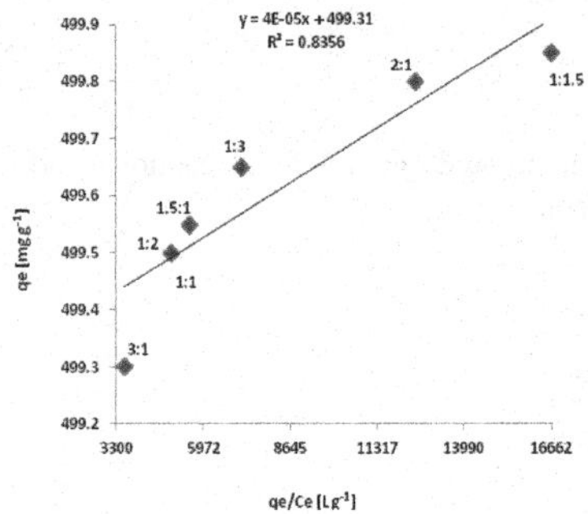

Figure 1. Langmuir adsorption isotherm of Pb^{2+} with ACF with phosphate groups $\left(q_e \ vs \ {q_e}/{C_e}\right)$

Adsorption isotherm of Freundlich

Figure 2 shows that the ratio 1:3, 1.5:1, 2:1, 1:1 and 1:2 to uptake Pb^{2+} with ACF with phosphate groups have the best behavior, indicating the removal capacity with respect the concentration of adsorbate in the heterogeneity of the fiber surface, on the other hand, the table I shows the results of the Langmuir and Freundlich adsorption isotherms.

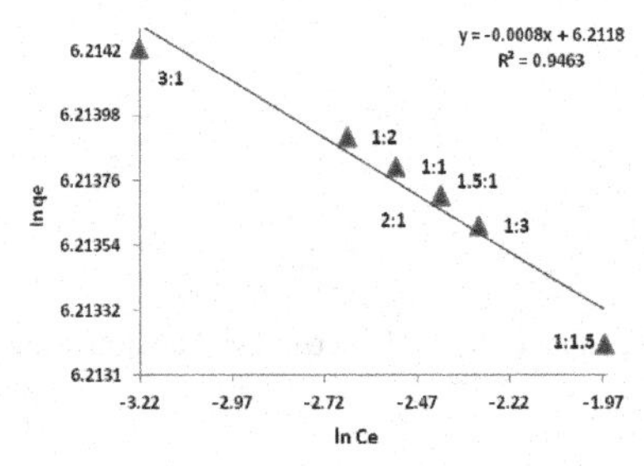

Figure 2. Freundlich adsorption isotherm of ACF with phosphate groups in the removal of Pb^{2+} $\left(\ln q_e \ vs \ \ln C_e\right)$

Table I. Langmuir and Freundlich adsorption isotherms of lead ions.

Langmuir adsorption isotherm	q_m [mg g^{-1}]	b [L mg^{-1}]	R^2 [%]	
	2.0E-03	12487512.59	0.84	
Freundlich adsorption isotherm	k_f	ln k_f	1/n	R^2 [%]
	6.2118	1.8264	-0.0008	0.94

Adsorption kinetic of pseudo-first order

For adsorption efficiency models are used kinetic pseudo-first order and pseudo-second order. The graph of figure 3 shows the results that included in table II where the rate of reaction of samples with ratio 1:3, 1.5:1 and 3:1 exhibit an elevated concentration time.

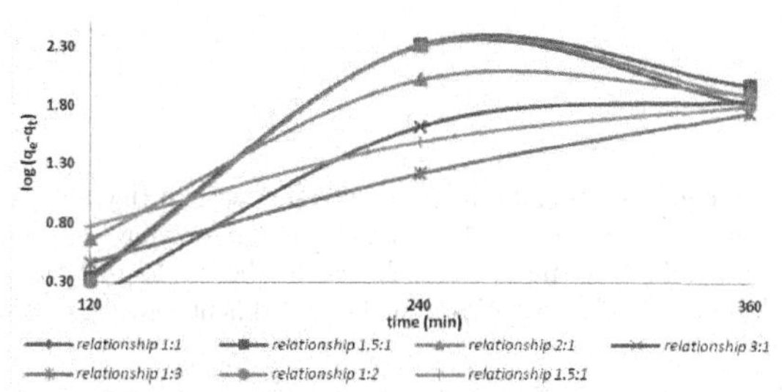

Figure 3. Model pseudo-first order of Pb^{2+} on ACF with phosphate groups $\left[\log(q_e - q_t) vs\ time\right]$

Adsorption kinetic of pseudo-second order

The graph in figure 4 is obtained as a result applying the model of pseudo-second order; all samples analyzed with this model give a better correlation because the reaction rate is proportional to the square of the concentration equilibrium, on the other hand, the table II shows the results of analysis of the kinetics of adsorption of Pb^{2+} using the models of pseudo-first order and pseudo-second order.

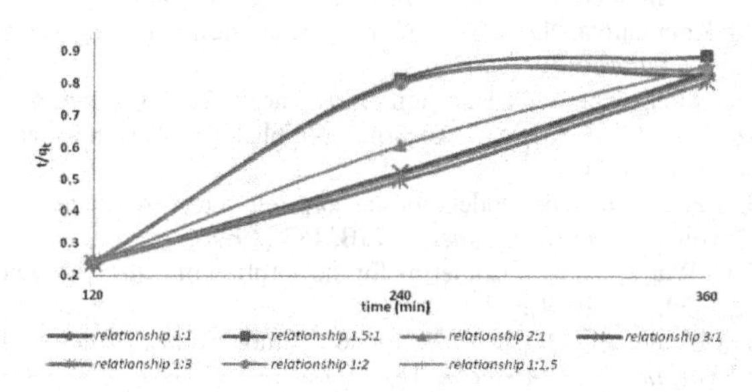

Figure 4. Model pseudo-second order of Pb^{2+} with ACF with phosphate groups $\left(t/q_t \text{ vs time} \right)$

Table II. Adsorption kinetic models Pb^{2+}

Relation	Model of pseudo-first order			Model of pseudo-second order			
	K_{ad} [min^{-1}]	q_e [$mg\ g^{-1}$]	R^2	q_e [$mg\ g^{-1}$]	h [$mg\ g^{-1}\ min^{-1}$]	k [$g\ mg^{-1} min^{-1}$]	R^2
1:1	0.0140	1.0359	0.52	400	2.89	1.81E-05	0.78
1.5:1	0.0159	0.7809	0.60	370	1428.571	0.010414286	0.83
2:1	0.0120	1.9024	0.68	385	-20.121	-0.000136016	0.99
3:1	0.0165	0.3281	0.85	400	-15.723	-9.83E-05	0.99
1:3	0.0124	0.7127	0.99	417	-18.657	-0.000107463	0.99
1:2	0.0152	0.8234	0.56	400	47.393	0.0002962	0.81
1.5:1	0.0099	2.0980	0.95	417	17.007	9.80E-05	0.99

CONCLUSIONS

The adsorption equilibrium and kinetic of activated carbon fibers with phosphate groups, derived from poly(acrylonitrile) fibers in the removal of lead ions have been studied. It has been found that adsorption equilibrium may be established by the model of Freundlich isotherm. On the other hand, the kinetic model of pseudo-second order indicates the best fit correlation.

REFERENCES

1. I. D. Harry, B. Saha, I. W. Cumming, "Effect of electrochemical oxidation of activated carbon fiber on competitive and noncompetitive sorption of trace toxic metal ions from aqueous solution" *J. of Colloid and Interface Sci.* **304,** (2010).
2. I. Langmuir, "The constitution and fundamental properties of solids and liquids" *J. Am. Chem. Soc.* **38** (1916).
3. X. Domémech, J. Paral, "Química ambiental de sistemas terrestres" *Ed. Reverté S.A., Barcelona Esp.* (2006).
4. W. C. Liu, H. H.Chem, W. H. Hsich, C. H.Chang, "Linking watersheed and eutrophication modelling for the shihmen reservoir" *Taiwan.* **54,** (11-12) (2006).
5. S. J. Allen, B. Koumaniva, "Decolourisation of water/istewater using adsorption" *J. Univ Chem Technol Metallurgy.* **40, (**2003).
6. H. M. F. Freundlich, "Über die adsorption in lösungen" *Z. Phys. Chem,* **57, (**1906).
7. M. Ahmaruzzaman, D. K. Sharma, "Adsorption of phenols from istewater" *J. of colloid and Interface Sci.* **287,** (2005).
8. Y. S. Ho, G. McKay, "Kinetic models for the sorption of dye from aqueous solution by wood" *Institution of Chemical Engineers,* **76B,** 183 (1998).
9. Y. S, Ho, C. C. Wang. "Pseudo isotherms for the sorption of cadmium ion onto tree fern" *Process Biotec.* **39,** 759(2004**).**
10. Y. S. Ho, G. McKay, D. A. J. Iste, C. F. Forster, "Study of the sorption of divalent metal ions on to peat" *Adsorption Sci. and Techn.* **18,** (2000).

Mater. Res. Soc. Symp. Proc. Vol. 1485 © 2013 Materials Research Society
DOI: 10.1557/opl.2013.209

Finite Element Simulation of Steel Quench Distortion- Parametric Analysis of Processing Variables

F. A. García-Pastor, R.D. López-García, E. Alfaro-López, M. J. Castro-Román
1 Cinvestav Unidad Saltillo, Carretera Saltillo-Monterrey Km 13, Ramos Arizpe, Coahuila, Mexico
[2] San Luis Rassini, Puerto Arturo No. 803, Piedras Negras, Coahuila, Mexico

ABSTRACT

Steel quenching from the austenite region is a widely used industrial process to increase strength and hardness through the martensitic transformation. It is well known, however, that it is very likely that macroscopic distortion occurs during the quenching process. This distortion is caused by the rapidly varying internal stress fields, which may change sign between tension and compression several times during quenching. If the maximum internal stress is greater than the yield stress at given processing temperature, plastic deformation will occur and, depending on its magnitude, macroscopic distortion may become apparent.

The complex interaction between thermal contraction and the expansion resulting from the martensitic transformation is behind the sign changes in the internal stress fields. Variations in the steel composition and cooling rate will result in a number of different paths, which the internal stresses will follow during processing. Depending on the route followed, the martensitic transformation may hinder the thermal stresses evolution to the point where the stress fields throughout the component may actually be reverted. A different path may support the thermal stresses evolution further increasing their magnitude. The cross-sectional area also affects the internal stresses magnitude, since smaller areas will have further trouble to accommodate stress, thus increasing the distortion. Additionally, the bainitic transformation occurring during relatively slow cooling rates may have an important effect in the final stress field state.

A finite-element (FE) model of steel quenching has been developed in the DEFORM 3D simulation environment. This model has taken into account the kinetics of both austenite-bainite and austenite-martensite transformations in a simplified leaf spring geometry. The results are discussed in terms of the optimal processing parameters obtained by the simulation against the limitations in current industrial practice.

INTRODUCTION

Quenching from the austenite phase field to improve steel hardness and strength through the martensitic transformation is a widely used process. However, macroscopic distortion is a common problem, which can appear during such processing. Macroscopic distortion is a result of the complex and rapidly variable stress field inside the component. There are two types of internal stresses during quenching: thermal and transformation-related stresses [1].

Thermal stresses are the result of contraction due to the rapidly decreasing temperature. The surface temperature drops faster than the core temperature, thus creating a compression-tension stress system which is eventually reverted, once the core temperature is comparable to the surface temperature. Transformation stresses, on the other hand, are the result of the martensitic transformation. This difussionless phase transformation yields a distortion along the *c* axis of the ferrite BCC cell, thus generating the BCT cell of the martensite phase. This transformation occurs at relatively low temperatures (~250-300°C) and as such, the transformation stresses are also a result of the temperature difference between the surface and the core. As expected, this situation leads to complex interaction between the two stress types, which are summarized in figure 1.

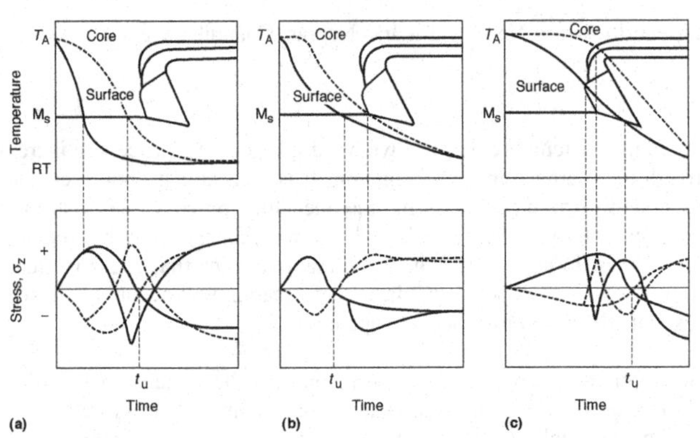

Figure 1. Internal stresses evolution during quenching, solid lines represent the evolution in the core, while dashes lines show the changes in the surface. The figure shows the temperature changes superimposed to CCT diagrams and their effect in the internal stress fields, ranging from fast (a), moderate (b) and slow (c) cooling [1]. For a more detailed explanation, please see text.

Figure 1 shows the temperature profiles plotted on top of CCT diagrams (top row), and their effect in the internal stress fields (bottom row). In the first case (figure 1a), it can be seen that if the martensitic transformation occurs before the thermal stresses change sign, the original stress field is reverted. This gives rise to a final stress field with the surface in tension and the core in compression. However, if the cooling rate is not as high as in the first case and the martensitic transformation occurs after the thermal stresses reversal (shown in figure 1b). Then the transformation stresses actually reinforce the previous stress state and further enhance the core to be in compression and the surface to be in tension. In the first two cases, the transformation is only from austenite to martensite. In the last case (figure 1c), the cooling rate is such that additional phases such as bainite or even ferrite and cementite are formed. The final stress field will thus depend on additional transformation to the martensitic phase. It is because of the different outcomes that may occur that proper transformation kinetics are needed if a quenching model is going to be developed.

THEORY AND MODELING

Common FE software can readily simulate transient heat transfer, even in complex geometries. However, in order to develop a proper quenching distortion model, it is necessary to include stress calculations, including both thermal and transformation stresses. For the simulations presented in this paper, the Deform 3D simulation software is used. This software is able to solve both the three-dimensional transient heat transfer equation, and the stress equations. Additionally, it is possible to assess the macroscopic distortion through a deformable mesh. Plasticity is taken into account using the generalized Johnson Cook model, using flow stress data from the extensive database provided by the software. A detailed review of the equation used in the model can be found in the work of Pietzch [2].

As shown in the previous section, it is fundamental to obtain proper transformation kinetics in order to develop a successful quenching model [3,4]. Current FE software such as Deform 3D or ABAQUS have extensive data bases which include mechanical, thermal and transformation properties for a wide range of materials. However, it has been found that the transformation data in this database do not account for a variable austenitic grain size. In fact, there is no information about the austenitic grain size for which the transformation kinetics are set. There is a number of recent efforts to improve the transformation kinetics used in this kind of distortion quenching models, mostly by dilatometry [5-7].

Transformation kinetics

In this work, temperature-time-transformation (TTT) curves for the bainitic transformation are obtained for three different austenitic grain sizes of SAE 5160 steel. A Linseis RITA (Rapid Induction Thermal Analysis) dilatometry equipment is used to analyze the length changes of cylindrical specimens, 5 mm in diameter and 10 mm long. For each one of the austenitic grain sizes, several isothermal transformations are carried out at different temperatures. Using this technique it is possible to construct three TTT diagrams for grain sizes 6, 8 and 10 ASTM. These diagrams are shown in figure 2.

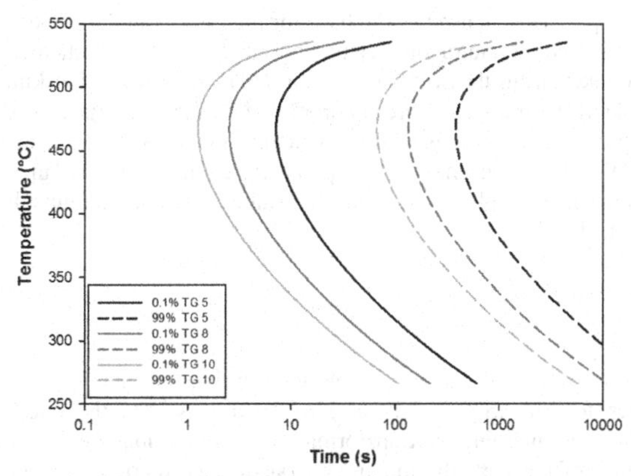

Figure 2. Effect of austenitic grain size in bainite transformation kinetics for SAE 5160 steel.

It is evident that a very high cooling rate is needed to avoid the formation of bainite during cooling of a steel sample with an ASTM size of 10.

Figure 3 shows the linear expansion during cooling of a SAE 5160 steel specimen with an austenitic grain size of 10 ASTM. It can be seen that there are no early deviations from the linear response when the cooling rate is above 30°C/s. However, if the cooling rate is considerably slowed down to 10°C/s, there is an deviation at approximately 450°C. This is attributed to the formation of bainite, following the TTT diagrams presented in figure 2. Figure 3 also shows that the martensite start temperature is slightly lower if bainite is previously formed during cooling. This effect has previously been reported in plain medium carbon steel. [6]

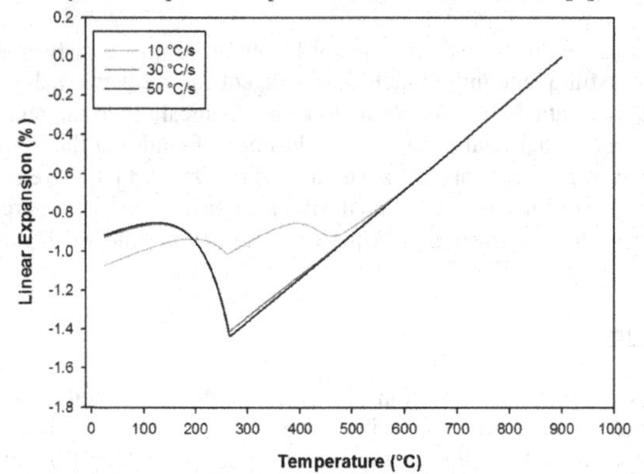

Figure 3. Effect of austenitic grain size in the linear expansion due to the bainite formation in SAE 5160 steel with a grain size of 10 ASTM.

FE modeling

A simplified leaf spring geometry is used as the base for the FE model developed in Deform 3D. The mesh is constructed by 40,000 tetragonal elements. The SAE 5160 mechanical, physical and thermal properties are taken from the software database. The transformation kinetics described previously for the ASTM 10 grain size are incorporated in the new material definition. The model considers the part being submerged in a quenching media with a constant heat transfer coefficient of 2,000 $Wm^{-2}K^{-1}$. The ends of the parts are submerged first, in a configuration commonly used in practice. A coupled thermo-mechanical and transformation solution is carried out as described by Pietzch [2].

DISCUSSION

Figure 4 shows the martensite volume evolution during quenching is finished. The transformation kinetics for an ASTM grain size of 10 are used in this model. It becomes immediately clear that the martensitic transformation is not complete by the end of the quenching. In the center of the part, the maximum transformed martensite fraction is close to 70%. The remaining 30% in the center of the part is transformed to bainite.

In spite of the martensitic transformation not being complete, it is possible to observe the macroscopic distortion due to the volume changes from austenite to ferrite. Figure 5 shows the displacement vectors during quenching. It can be seen that the martensite formation generates a vector field, which distorts the part. In this particular geometry and immersion sequence, the two martensitic fronts come from each end and find themselves at the center of the part. Once the fronts interact with each other, it is very difficult for the part to accommodate all the stresses elastically, which in turn increases the likelihood of plastic distortion.

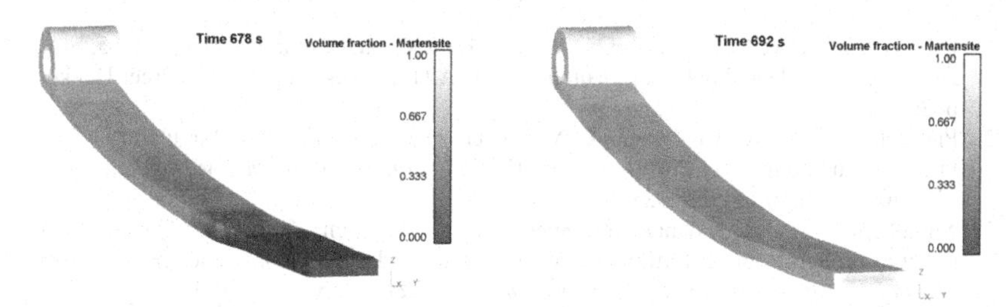

Figure 4. Martensite volume fraction during (left-hand side) and after (right-hand side) quenching. For clarity, only half of the computational model is shown and the part is rotated 180° around the x axis.

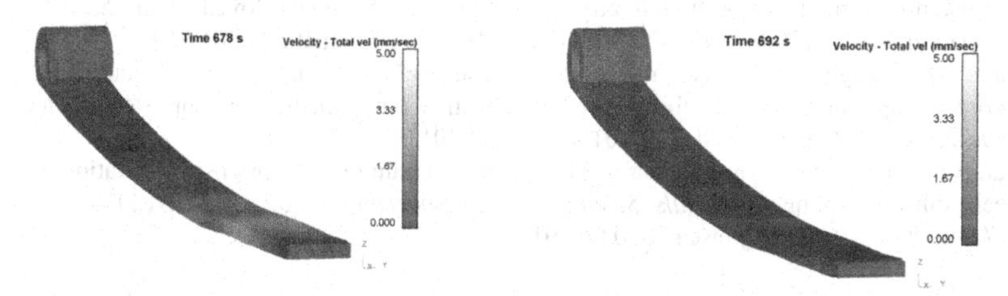

Figure 5. Velocity contours during (left-hand side) and after (right-hand side) quenching. Notice the displacement occurring during the martensitic transformation. Once the quenching is finished, there are no displacements in the component. For clarity, only half of the computational model is shown and the part is rotated 180° around the x axis.

CONCLUSIONS

It has been shown that it is possible to successfully represent the distortion during quenching in a leaf spring made of SAE 5160 steel using transformation kinetics specifically determined for a particular austenitic grain size. The results show that small changes in the austenitic grain size have an important effect in the critical cooling rate to avoid bainite formation. This improved transformation kinetics allows a proper representation of the distortion phenomena during steel quenching. However, a more thorough model needs to include a variable austenitic grain size through thickness. This variable austenitic grain size gains importance in thicker cross-sections, in which steeper temperature gradients can be found. Finally, it is important to mention that it is found that previous diffusion transformations have the effect of depressing the martensite start temperature. Such effect needs to be addressed in the FE modeling in order to predict distortion accurately.

REFERENCES

1. G.E. Totten and M.A.H. Howes, Distortion of Heat Treated Components, chapter 5, Steel Heat Treatment Handbook, G.E. Totten and M.A.H. Howes, Ed., 1997, Marcel Dekker, p. 292.
2. Pietzsch, R., Brzoza, M., Kaymak, Y., Specht, E., & Bertram, A. Simulation of the Distortion of Long Steel Profiles During Cooling. *Journal of Applied Mechanics*, *74*(3), 427 (2007). doi:10.1115/1.2338050
3. Nallathambi, A. K., Kaymak, Y., Specht, E., & Bertram, A. (2010). Sensitivity of material properties on distortion and residual stresses during metal quenching processes. *Journal of Materials Processing Technology*, *210*(2), 204–211. doi:10.1016/j.jmatprotec.2009.09.001
4. S. J. Lee, Y. K. Lee. Finite element simulation of quench distortion in a low alloy steel incorporating transformation kinetics. *Acta Materialia* No. 56 (2008), p. 1482-1490. doi:10.1016/j.actamat.2007.11.039
5. San Martín, D., Rivera-Díaz-del-Castillo, P. E. J., & García-de-Andrés, C. In situ study of austenite formation by dilatometry in a low carbon microalloyed steel. *Scripta Materialia*, No. 58 (2008)., pp. 926–929. doi:10.1016/j.scriptamat.2008.01.019
6. Jung, M., Kang, M., & Lee, Y.-K. (2012). Finite-element simulation of quenching incorporating improved transformation kinetics in a plain medium-carbon steel. *Acta Materialia*, *60*(2), 525–536. doi:10.1016/j.actamat.2011.10.007
7. Van Bohemen, S. M. C., & Sietsma, J. The kinetics of bainite and martensite formation in steels during cooling. *Materials Science and Engineering: A*, *527*(24-25) (2010), pp 6672–6676. doi:10.1016/j.msea.2010.06.091

Mater. Res. Soc. Symp. Proc. Vol. 1485 © 2013 Materials Research Society
DOI: 10.1557/opl.2013.210

3D Computational Simulation of Multi-Impact Shot Peening

Juan Solórzano-López and Francisco Alfredo García-Pastor
CINVESTAV Saltillo, Coahuila, México.

ABSTRACT

Shot peening is a widely applied surface treatment in a number of manufacturing processes in several industries including automotive, mechanical and aeronautical. This surface treatment is used with the aim of increasing surface toughness and extending fatigue life. The increased performance during fatigue testing of the peened components is mainly the result of the sub-surface compressive residual stress field resulting from the plastic deformation of the surface layers of the target material, caused by the high-velocity impact of the shot. This compressive residual stress field hinders the propagation and coalescence of cracks during the second stage of fatigue testing, effectively increasing the fatigue life well beyond the expected life of a non-peened component.

This paper describes a 3D computational model of spherical projectiles impacting simultaneously upon a flat surface. The multi-impact model was developed in ABAQUS/Explicit using finite element method (FEM) and taking into account controlling parameters such as the velocity of the projectiles, their incidence angle and different impact locations in the target surface. Additionally, a parametric study of the physical properties of the target material was carried out in order to assess the effect of temperature on the residual stress field.

The simulation has been able to successfully represent a multi-impact processing scenario, showing the indentation caused by each individual shot, as well as the residual stress field for each impact and the interaction between each one of them. It has been found that there is a beneficial effect on the residual stress field magnitude when shot peening is carried out at a relatively high temperature. The results are discussed in terms of the current shot-peening practice in the local industry and the leading edge developments of new peening technologies. Finally, an improved and affordable processing route to increase the fatigue life of automotive components is suggested.

INTRODUCTION

Shot peening process is widely applied industrial surface treatment in several manufacturing industries, such as automotive, mechanical and aeronautical. The objective of this treatment is to increase surface toughness and extend the fatigue life of mechanical pieces. This improved performance of the peened components is the result of the sub-surface compressive residual stress field resulting from the plastic deformation of the surface layers of the target material, caused by the high-velocity impact of the shot [1-5]. This compressive residual stress field hinders the propagation and coalescence of cracks during the second stage of fatigue testing, effectively increasing the fatigue life well beyond the expected life of a non-peened component. Evaluation of residual stresses field is realized by experimental methods like X-ray diffraction. This method is both destructive and costly in time and money and the needed skills for its realization are considerable. For this reason, numerical simulation of shot peening process is an alternative in determination of residual stress, simulations can be made varying all the inherent process parameters, including velocity, size and impact angle of the projectiles. Additionally, it is possible to carry out simulations which take into account conditions such as an elevated temperature of the target or the assess presence of a stress field due a previous deformation. The high speed of shot process is simulated using explicit dynamics method, because the load is applied rapidly and therefore the mechanical response of the changes in the target and projectiles is almost instantaneous [1-5].

The explicit analysis in Finite Element Modeling solves the problem by analyzing the motion of stress and deformation waves through the nodes of the structure using small time increments [1].

The explicit algorithm in ABAQUS adopts a central difference rule to integrate the equations of motion explicitly through time, using kinematic conditions at each increment to calculate the kinematic conditions for the next increment. Additionally, master-slave algorithm was used to enforce the contact constraints, the rigid surface of the spherical projectiles is chosen as master contact surface.

THEORY

Numerical simulation is a useful tool to study and analysis of results of several mechanical processes. In the present work, a 3D numerical model was developed in ABAQUS commercial FE software. The aim is explore the shot peening process varying the impact angle, and size of them. Also, to obtain information of residual stress caused by shots impacting onto a target cube at 200°C. The target material was steel, Young´s modulus 200e9 Pa and Poisson´s ratio of 0.285. Projectiles were rigid spherical bodies, covered by an analytical rigid layer in order to accomplish the impact performance between two surfaces. As it was mentioned, the master surface is the projectile surface, and slave surface is the target body surface. Target body was discretized in 125 000 orthogonal nodes and projectile impacts were in 90° and 45°, velocities varied from 10 to 23 m/s and diameters used were 1 and 0.3 mm.

Software solved several equations, such as the Motion Equation and its auxiliary equations (velocity and acceleration equations), Time Increment (since explicit method is being used) and Johnson-Cook empirical model and its parameters. Also, relationship between hardness and yield strength, hydrodynamic pressure developed during the impact and Internal Energy as temperature function (see table I).

Table I. Solved equations using ABAQUS commercial software [1].

Equation	Description
$\dot{u}^{(i+\frac{1}{2})} = \dot{u}^{(i-\frac{1}{2})} + \frac{\Delta t^{(i+1)} + \Delta t^{(i)}}{2} \ddot{u}^{(i)}$ $u^{(i+1)} = u^{(i)} + \Delta t^{(i+1)} \dot{u}^{(i+\frac{1}{2})}$	Motion Equation
$\dot{u}^{(i+1)} = \dot{u}^{(i+\frac{1}{2})} + \frac{1}{2} \Delta t^{(i+1)} \ddot{u}^{(i+1)}$	Linear interpolation of mean velocity
$\dot{u}^{(+\frac{1}{2})} = \dot{u}^{(0)} + \frac{\Delta t^{(1)}}{2} \ddot{u}^{(0)}$	Mean velocity for first increment

$\ddot{u}^{(i)} = M^{-1}(F^{(i)} - I^{(i)})$	Acceleration at the beginning of time increment
$\Delta t \leq \frac{2}{\omega_{max}}$	Time increment
$\sigma_y = (\varepsilon_p, \dot{\varepsilon}, T) = \sigma_0 \left[1 + \frac{B}{\sigma_0}(\varepsilon_p)^n\right]\left[1 + C\ln(\dot{\varepsilon}^*)\right]\left[1 - (T^*)^m\right]$	Johnson-Cook Model
$\dot{\varepsilon}^* = \frac{\dot{\varepsilon}}{\dot{\varepsilon}_0}$ $T^* = \frac{(T - T_r)}{(T_m - T)_r}$	Johnson-Cook Model Parameters
$\sigma_0 = \exp(A_1 R_C + A_2)$	Hardness-Yield Stress Relation
$p = \frac{\rho_0 C_0^2(\eta - 1)\left[\eta - \frac{\Gamma_0}{2}(\eta - 1)\right]}{\left[\eta - S_a(\eta - 1)\right]^2} + \Gamma_0 E$ $\eta = \frac{\rho}{\rho_0}$	Hydrodynamic pressure developed during the impact
$E = \frac{1}{V_0}\int C_V dT \approx \frac{C_V(T - T_0)}{V_0}$	Internal Energy as Temperature Function

DISCUSSION

The results of simulations are shown in the next figures, which synthetize information about residual stress fields due projectiles impact and plastic surface deformation. Figure 1 shows the studied system, cubic target body (1 cm^3) and projectiles.

Figure 1. Geometry of studied system showing the target body discretization and 5 projectiles.

Figures 2 and 3 are from the same impact result, the figure 2 shows simply the resulting residual stress field of an impact of 1 mm diameter projectile traveling at 23 m/s and impacting the target surface in a 45° angle. Figure 3 shows the linear profile of residual stress obtaining technique, as ABAQUS display.

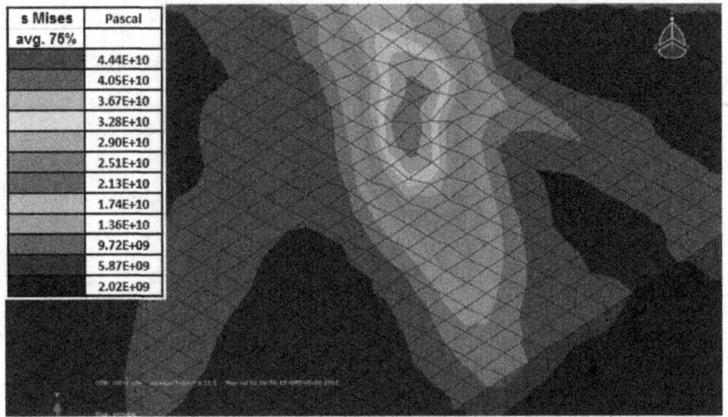

Figure 2. Residual stress field due a 1 mm projectile impacting at 23 m/s and 45°.

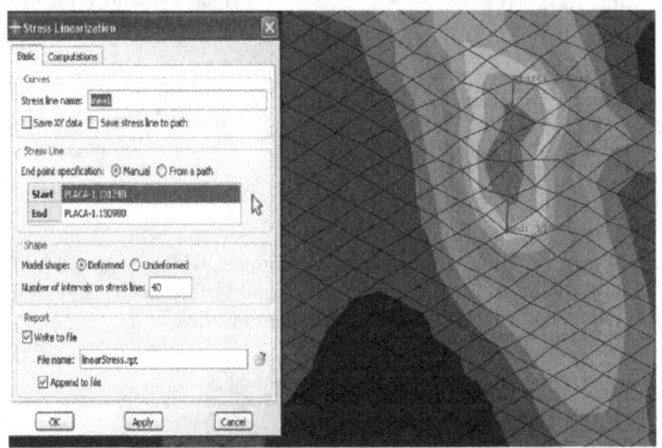

Figure 3. Same impact of figure 2 showing the linear profile of residual stress obtaining technique.

Figure 4 shows an impact dimple of a 0.3 mm diameter projectile that impact at 90° and 10 m/s velocity, and the graphic of linear residual stress profile data obtained with ABAQUS.

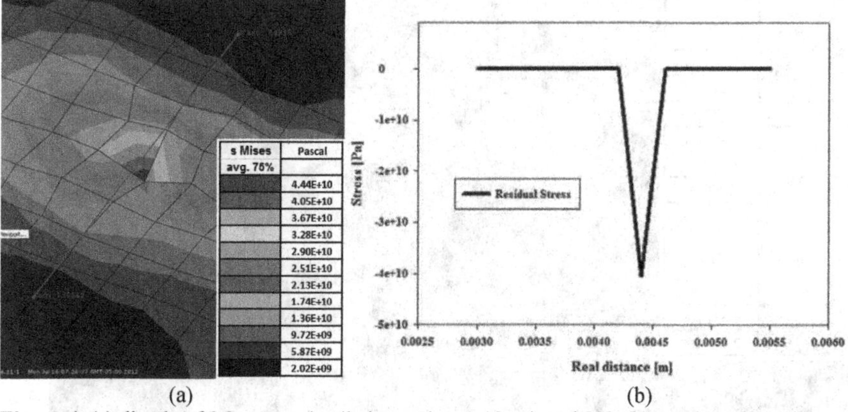

Figure 4. (a) dimple of 0.3 mm projectile impacting at 10 m/s and 90°. (b) graphic of linear residual stress profile.

Figure 5 shows two impacts in the same body target, in the top of the figure is the dimple of 1 mm diameter projectile that impacts at 23 m/s a 45°, and in the bottom, the dimple of a 0.3 mm hitting the surface at 90° and 10 m/s.

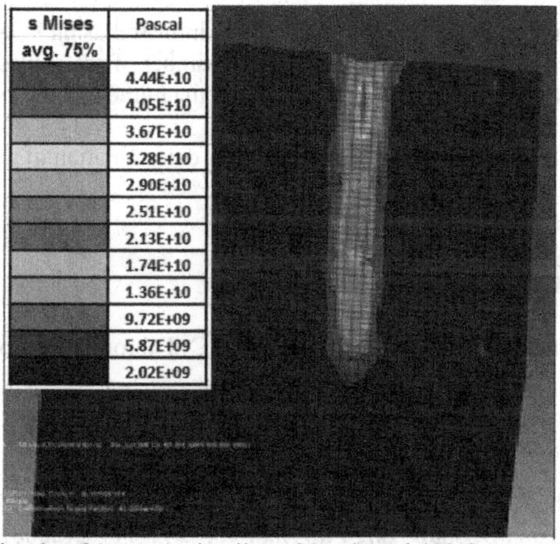

Figure 5. In the top, dimple of 1 mm projectile at 23 m/s and 45°. Lower dimple of a 0.3 mm projectile impacting at 10 m/s and 90°.

The next step of this work is to substitute the spherical projectiles by cylindrical projectiles. The cause of this is the use of cut wire low cost munitions instead spherical cast (and comparative expensive) munitions. Figure 6 shows preliminary results of this effort.

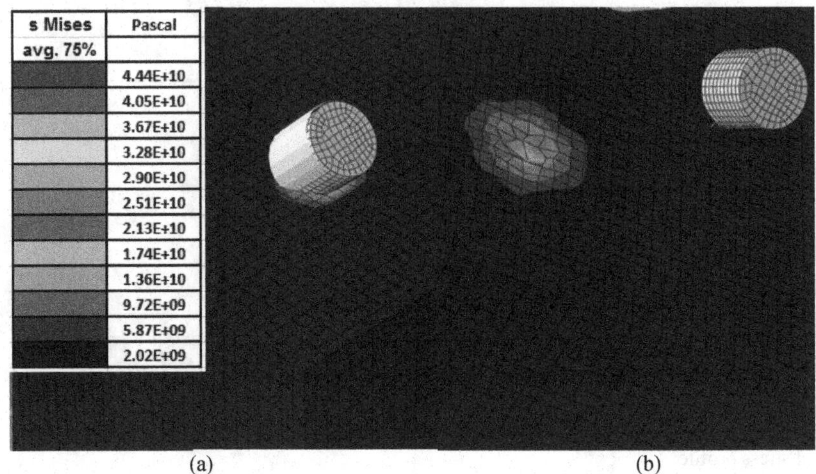

Figure 6. (a) Impact of a cut wire projectile and, (b) resulting dimple and residual stress field.

CONCLUSIONS

- The shot peening process can be satisfactorily simulated using FE software varying conditions such as impact angle, velocity and number of projectiles.
- The residual stress fields obtained have similar magnitude factor that experimental or industrial practice data.
- Mathematical modeling is an important tool for determination of results at low cost than experiment performance.

ACKNOWLEDGMENTS

The authors acknowledge the funding from CONACyT through Dr. Solórzano Postdoc.

REFERENCES

1. R. Shivpuri, X. Cheng and Y. Mao, *Mater Design* **30**, 3112-3120, (2009).
2. T. Hong, J. Y. Ooi and B. Shaw, *Eng Fail Anal* **15**, 1097-1110, (2008).
3. T. Kim, J. H. Lee, H. Lee and S. Cheong, *Mater Design* **31**, 50-59, (2010).
4. M. Frija, T. Hassine, R. Fathallah, C. Bouraoui and A. Dogui, *Mat Sci Eng A-Struct* **426**, 173-180 (2006).
5. G. H. Majzoobi, R. Azizi and A. A. Nia, *J Mater Process Tech* **164-165**, 1226-1234, (2005).

Mater. Res. Soc. Symp. Proc. Vol. 1485 © 2013 Materials Research Society
DOI: 10.1557/opl.2013.211

Study of Aluminum Degasification with Impeller-Injector Assisted by Physical Modeling

M. Hernández-Hernández[1], E. A. Ramos-Gómez[1], M. A. Ramírez-Argáez[1]
[1]Facultad de Química, UNAM, Departamento de Ingeniería Metalúrgica. Edificio "D" Circuito de los Institutos s/n, Col. Cd. Universitaria, C.P. 04510 México D.F., México.

ABSTRACT

A full-scale water physical model of a degassing unit is built and used to evaluate the performance of several impeller designs. Four impeller designs are tested: a) one smooth not commercial impeller for reference purposes, b) a commercial design by FOSECO®, called standard impeller in this work, c) a commercial design by FOSECO® with notches, and d) a new design proposed in this work. Since the physical model is easy and safe to operate, a full experimental design is performed to evaluate the effect of the most important process variables, such as impeller rotating speed, gas flow rate, impeller design and the point of gas injection (a conventional gas injection through the shaft and a novel method of injecting gas through the bottom of the ladle) on the kinetics of oxygen desorption of water which is similar to dehydrogenation of liquid aluminum. The new design of impeller proposed in this work shows the best performance in degassing of all impellers tested in this study. It is found that the rotor speed and its design are the most significant variables affecting degassing kinetics, and therefore the analysis of the existing commercial impeller designs may be useful to optimize the fluid dynamics of the process, which in turn would increase efficiency and productivity of the process. Finally, the novel gas injection method through the bottom, proposed by our own group, presents slightly faster degassing kinetics than the conventional injection of purge gas in the conventional way through the impeller.

INTRODUCTION

Aluminum alloys are used in several applications in automotive, aerospace and food technology industries, among others. The versatility of those materials is due to a combination of low density and high mechanical resistance. However, molten aluminum is susceptible to dissolve hydrogen from atmospheric humidity. Hydrogen solubility increases proportionally with an increase in temperature but decreases during aluminum solidification, resulting in parts with porosities, which are detrimental to the mechanical properties of the manufactured castings [1-3]. The actual industrial technology for liquid aluminum dehydrogenation consists in purging gases into the liquid aluminum alloy during a vigorous agitation by a graphite impeller rotating at high-speed velocities. The gas is conventionally injected through the rotor shaft. Due to the shear of the rotor, the gas is dispersed in the form of small bubbles in order to increase the degassing efficiency. Several variables can be identified in this process: gas flow rate, geometric impeller design, impeller rotating speed, and gas injection position, among others. Optimal degassing parameters can be obtained by performing experiments varying the most important variables already mentioned. Since the implementation of an experimental design is difficult and costly to achieve in the industrial facilities, a rigorous physical model, based on similarity criteria (geometrical, dynamic, kinematic, etc.), is a good analysis tool to understand, control and optimize this process [4-11]. In this work, a full-scale water physical model is built with transparent acrylic to simulate an industrial degassing ladle. Also, the experimental results obtained from this water model, are used to propose degassing kinetics equations that fit our data through a global mass transfer coefficient in order to describe the degasification kinetics with a few parameters. Besides, these experimental results from our physical model will be used to validate a two phase fluid flow mathematical model under construction.

EXPERIMENT

The full-scale water physical model is built in transparent acrylic, and filled with distilled water to simulate molten aluminum. The shaft and four different impellers (different designs) are made of Nylamid®. Before degassing, water is saturated with oxygen by injecting air through a commercial compressor (saturation point is reached around 6 ppm of dissolved oxygen at room temperature). Once saturated with oxygen, the bath is deoxidated by injecting nitrogen (99.99% N_2 at gas flow rates of 10 and 40 l/min). Four impellers are tested: a) smooth design. b) standard design, c) commercial design with notches, and d) a new proposed impeller design.

These impellers with different geometries are tested in order to determine the best design for fastest oxygen removal rate, which is similar to dehydrogenation of liquid aluminum.

The gas is injected through the impeller shaft (conventional industry practice) or at the bottom of the vessel (novel injection). Two values of impeller's rotation speed are considered, 290 and 580 RPM. The oxygen concentration is measured using a commercial oxymeter. Each experiment is performed by triplicate. The experimental set up is described in Figure 1, where the two gas injection points tested are graphically represented. Figure 1 also shows the impellers geometries.

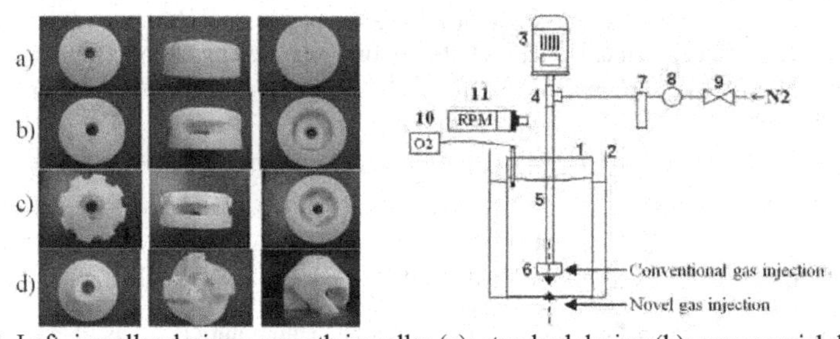

Figure 1. Left, impeller designs: smooth impeller (a); standard design (b); commercial design with notches (c); new design proposed (d). Right, scheme of the experimental setup: 1 acrylic vessel, 2 acrylic container, 3 motor, 4 rotary joint, 5 shaft, 6 impeller, 7 flowmeter, 8 pressure regulator, 9 valve, 10 oxymeter, and 11 tachometer.

RESULTS AND DISCUSSION

Physical water model results

Figure 2 shows the dimensionless concentration of dissolved oxygen (instantaneous oxygen concentration $[O_2]$ divided by the initial oxygen concentration $[O_2^\circ]$) in the water bulk as a function of time. Four different impeller designs are tested; Smooth (Figure 2a), Standard

(Figure 2b), Commercial with notches (Figure 2c) and the New Design (Figure 2d). As a reference, experiments with the smooth impeller design are conducted, but the agitation produced is the worst. Additionally, the size and distribution of nitrogen bubbles did not perform well for degassing because bubbles are big and concentrated in a small region in the center of the ladle (see Figure 3a). The lowest degassing time reached with this smooth impeller are 1065 seconds, using 40 l/min gas flow rate and 580 RPM with both gas injection techniques, but in the rest of the cases with this smooth impeller, degasification is not completed and the oxygen concentration is asymptotic at long times (see Figure 2a).

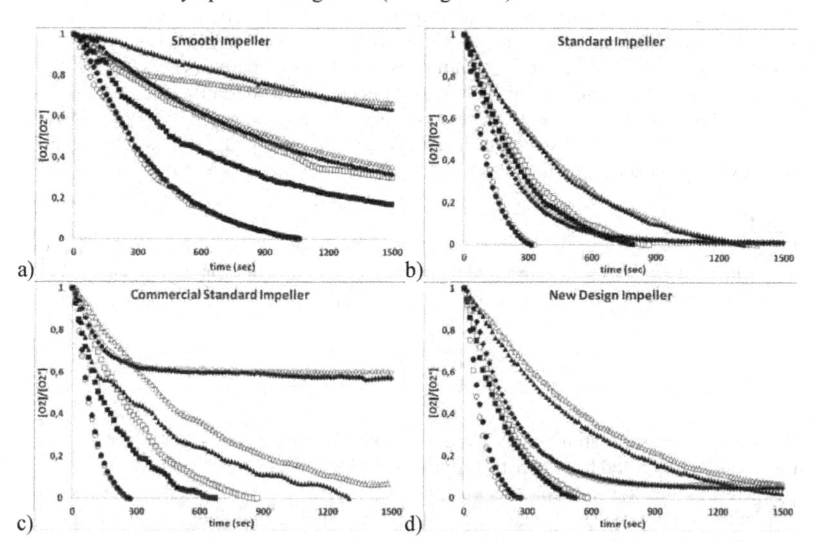

Figure 2. Dimensionless concentration of dissolved oxygen as a function of time during the degassing process. Impellers tested are: Smooth (a), Standard (b), Commercial with notches (c) and the New Design proposed in this work (d). Conventional and Novel gas injection are white and black symbols respectively. Conditions are: Δ 290 RPM and 10 L/min, □ 290 RPM and 40 L/min, ◇ 580 RPM and 10L/min, and ○ 580 RPM and 40 L/min.

The standard impeller (Figure 2b) shows the best degasification kinetics at 580 RPM and 40 l/min of gas flow rate, where complete oxygen desorption is reached in 330 seconds (conventional injection) and in 315 seconds (novel injection). In the other cases, with lower values of both angular speed and gas flow rate, degassing times are larger.

Commercial design impeller with notches (Figure 2c) produces degasification with the same general trend as in the previous geometrical design (standard impeller) but faster. At 580 RPM and 40 l/min, complete oxygen desorption is reached after 270 seconds (conventional and novel injections with the same kinetics). The better performance of this impeller (compared against the smooth) can be explained as the nitrogen big bubbles are broken and dispersed in the bath as small bubbles more efficiently with this design (see Figure 3c). The worst performance of degasification (longest times of degasification) are obtained by using gas flow rates of 10 l/min and rotation speeds of 290 RPM for all impellers except for the impeller with notches, which present the worst degassing conditions (oxygen desorption has an asymptotic trend) by using the highest value of impeller's velocity (580 RPM) with the minimum value of purging gas flow rate, which can be explained because the turbulence produced by agitation promotes the pick-up of atmospheric oxygen at the free surface canceling the action of the purging nitrogen bubbles. Best conditions for a fast degassing kinetics for all impellers designs tested are 580 RPM and 40 l/min. However, the excess in gas employed is not necessarily efficient in terms of purge gas consumption.

The proposed impeller with the new design (Figure 2d) is the most efficient in degasification kinetics of all designs of the impellers tested. At 40 l/min of gas flow rate and rotation speed of 580 RPM, the oxygen desorption time is completed at approximately 225 seconds (conventional injection), and 270 seconds (novel injection). The best efficiency of degasification with this impeller means that the proposed design is a good option in the hydrogen removal from the molten metal bulk, because the stirring degree of the liquid, the dispersion and size of the bubbles are optimal (see Figure 3d).

Degassing kinetics

Our experimental kinetic results are fitted using a first order kinetic equation (equation 1) to obtain a kinetic parameter (***k***) with a reasonable good correlation in all cases ($R^2 \geq 0.94$):

$$\left[\frac{O_2}{O_2^0}\right] = a + b\exp(-kt) \qquad (1)$$

where the dimensionless concentration of dissolved oxygen in the bulk $[O_2]/[O_2^\circ]$ is a function of time, t, in seconds, while a and b are constants and ***k*** is the kinetic parameter, which quantitatively measures the oxygen desorption kinetics.

First order kinetic parameters, ***k***, calculated for all experimental conditions tested in our work with equation (1) are presented in the Table I.

Figure 3. Water model photos for the different impellers tested, (a); standard design (b); commercial design with notches (c); new design proposed (d) at 580 RPM and gas flow rate of 40L/min. (Novel gas injection at the bottom of ladle).

Table I. Kinetic parameters obtained from the experimental results.

Exp. No.	N_2 Flow Rate, (l/min)	Rotor speed RPM	Point of Gas Injection	Kinetic Parameter, k [s^{-1}] x 10^3			
				Smooth Impeller	Standard Impeller	Commercial Impeller with notches	New Design Impeller
1	10	290	Convent.	2.2	1.96	1.68	1.51
2	40	290	Convent.	0.91	3.20	3.02	4.43
3	10	290	Novel	0.3	2.07	2.30	1.77
4	40	290	Novel	1.78	3.60	4.80	4.73
5	10	580	Convent.	0.94	4.98	8.44	5.08
6	40	580	Convent.	2.70	7.50	9.57	11.05
7	10	580	Novel	1	5.14	8.05	4.67
8	40	580	Novel	2.62	7.97	9.22	10.04

The observed variations in ***k*** values (see Table I) are the result of variations in impeller rotating speed, gas flow rate, and in a less extent in changes in the point of gas injection. It can be seen that there is a slightly difference between both gas injection types. The new injection point advantages the conventional injection because the former method may reuse graphite shafts and impellers with cracks or fractures, increasing the life of these components, saving materials and reducing the production costs in industrial systems. The new design impeller proposed in this work had the highest values of the kinetic parameter at high gas flow rates. At low flow rates, the impeller with notches is the best design. Then, the kinetic parameter ***k*** may be a good indicative for the evaluation of the degassing practice, since this number includes the mass transfer coefficient, the total surface area of bubbles and other important kinetic variables at specific conditions. However, these results show a complex and non linear relation between process variables and degassing kinetics.

CONCLUSIONS

Degasification times are determined in experiments performed in this work in a full-scale water physical model.

Degasification times depend on several variables, being the principal of them the impeller design. At the same rotation speed of the impeller and the same purging gas flow rate, degasification time decreases if the impeller is able to disperse gas in the form of small bubbles in the entire ladle that promotes an efficient degasification of the liquid.

The most efficient impeller geometry for degassing is the new design proposed in this work at high gas flow rates and the impeller with notches at low gas flow rate.

Impeller design with notches promotes pick-up of the atmospheric oxygen canceling the degassing effect of the purging gas bubbles at high rotor speeds.

Regarding the injection point techniques, the novel injection is slightly more efficient than the conventional (standard impeller), but with the impeller with notches, degassing time is practically the same for both injection points and in the case of the impeller with the new design proposed in this work, the conventional gas injection resulted to be more efficient.

The degassing kinetics is successfully described with a first order kinetics equation, being the kinetic parameter ***k*** being very sensitive to the impeller design but insensitive to the point of gas injection. This parameter involves the combined influence of the each process variable (injection point, gas flow rate, impeller rotating speed, and even geometry of the impeller).

In future work we will analyze the relative performances of the same impellers in a pilot plant with molten aluminum in order to validate the results from the physical model.

ACKNOWLEDGMENTS

M. Hernández-Hernández acknowledges financial support from CONACyT through a Ph.D. scholarship.

REFERENCES

1. G. K. Sigworth and T. A. Engh, *"Chemical and kinetic Factors Related to Hydrogen Removal from Aluminum"*, Metall. Trans. B, **13**, (1982).
2. J. G. Stevens and H. Yu. *Light Metals.* **121**, (1992), The Minerals, Metals & Materials Society (TMS), Warrendale, PA, USA.
3. J. Szekely, *"Fenómenos de Flujo de Fluidos en el Procesamiento de Metales"*. Limusa, México, (1988).
4. L. Zhang, X. Lv, A. T. Torgerson and M. Long, *"Removal of Impurity Elements from Molten Aluminum: A Review"*, Miner. Process. Extr. Metall. Rev., **32**, (2011).
5. J. Szekely, J. W. Evans and J. K. Brimacombe, *"The Mathematical and Physical Modeling of Primary Metals Processing Operations"*. John Wiley & Sons, USA, (1988).
6. F. Boeuf, M. Rey y E. Wuilloud, **"** *Metal batch tretment optimisation of rotor runnig condition"*, The Minerals, Metals & Materials Society, 1993, Vol. Light Metals 1993.
7. Thay Nilmani, " *A comparative study of impeller performance"***.** s.l. : The Minerals, Metals & Materials Society, 1992, Vol. Light Metals 1992.
8. Mariola Saternus and Jan Botor, **"** *The physical and mathematical model of aluminium refining process in reactor URO – 200"***.** USA : The Minerals, Metals & Materials Society, 2005.
9. Waldemar Bujalski, *"Mixing Studies Related to the cleaning of molten aluminium"* 3, Weinheim, Germany : WILEY-VCH, 2004, Vol. 27.
10. Mi Goufa," *Research on water simulatiion experiment of the rotating impeller degassing process",* China : Materials Science and Engineering A, 2008. 0921-5093.
11. J. L. Camacho-Martínez, M. A. Ramírez-Argáez and R. Zenit, *"Physical Modelling of Aluminium Degassing Operation with Rotating Impellers –A Comparative Hydrodynamic Analysis"*, Mater. Manuf. Processes, **25**, (2010).

Mater. Res. Soc. Symp. Proc. Vol. 1485 © 2013 Materials Research Society
DOI: 10.1557/opl.2013.212

Glass ceramic materials of the SiO_2-CaO-MgO-Al_2O_3 system: Structural characterization and fluorine effect

Mitzué Garza-García[1], Jorge López-Cuevas[2] and Oscar Hernández-Ibarra[3]
[1]Universidad Autónoma de Coahuila, Escuela Superior de Ingeniería, Blvd. Adolfo López Mateos s/n, C.P. 26800, Nueva Rosita, Coahuila, México.
[2]Centro de Investigación y de Estudios Avanzados del IPN, Unidad Saltillo, Carr. Saltillo-Monterrey, Km. 13.5, C.P. 25900, Ramos Arizpe, Coahuila, México.
[3]Instituto Tecnológico de La Región Carbonífera, Km 120, Carretera 57, Villa de Agujita, Municipio de Sabinas, Coahuila, México.

ABSTRACT

Glass-Ceramic monoliths of the SiO_2-CaO-MgO-Al_2O_3 system are obtained in this research. Due to its potential dual role as a flux and as a nucleating agent, two CaF_2 levels (X = 3 and X = 6 mol.%) are investigated in the parent glass composition. Due to its good mechanical properties, we intend to obtain Diopside-type pyroxene [(Ca)(Mg,Al)(Al,Si)$_2$O$_6$] as the main crystalline phase in the synthesized glass-ceramics. Vickers microhardness (HV), density and type of crystallization are determined in the latter materials. The morphology and size of the Diopside crystals, as well as the crystallized fraction, are determined with the help of Scanning Electron Microscopy (SEM) and X-Ray Diffraction (XRD). Both materials exhibit surface crystallization with Diopside-type pyroxene phase with acicular morphology homogeneously distributed in the glassy matrix. The specimen with the least amount of added fluorine shows the highest microhardness value, as well as the largest and thickest acicular crystals of Diopside-type pyroxene, the lowest apparent density and the largest crystallized fraction. Our results indicate that CaF_2 added in the amounts used by us does not act as nucleating agent, but it does affect the growth of the acicular crystals of the Diopside-type pyroxene phase. This is attributed mainly to the effect of fluorine on the glass structure and properties. The materials developed in this study may be considered as viable alternatives for applications in abrasive and corrosive environments, as well as for substrates for metallic coatings, and for abrasion-resistant floor tiles and other structural applications.

INTRODUCTION

Glass-ceramics typically contain between 50 and 90 vol.% of crystalline phases, the rest of the materials consists of a residual glassy phase. In comparison with many glasses and ceramics, the glass-ceramic materials have relatively higher mechanical and abrasion strength, as well as higher resistance to the attack by chemical reagents [1]. The properties of these materials depend on the size, morphology, chemical composition and lattice structure of their crystalline phases, as well as on the distribution of the latter in the glassy matrix and on the chemical composition of this phase [1,2].

D.U. Tulyaganov et al. [3] investigated glass-ceramic materials with Diopside-type pyroxene [(Ca)(Mg,Al)(Al,Si)$_2$O$_6$] as the main crystalline phase, plus other secondary crystalline phases. These authors mentioned that their materials, in addition to their good mechanical properties, are also biocompatible. It is known that the high twinning microstructural characteristic shown by Diopside, which constitutes an energy absorbing mechanism, produces a relatively high tensile strength in the glass-ceramic materials that are based on this phase [4].

The effect of some additives (B_2O_3, P_2O_5, Na_2O and CaF_2) on the properties of glass-ceramic materials of the CaO-MgO-SiO_2 system has been studied by D.U. Tulyaganov et al. [5]. These compounds acted mainly as fluxes. A. Rafferty et al. [6] studied the influence of fluorine content in Apatite ($Ca_{10}P_6O_{25}$)-Mullite ($Al_6Si_2O_{13}$) glass-ceramics. Fluorine acted in two ways: as a nucleating agent and as a facilitator for the kinetics of crystallization.

This investigation is aimed to study the effect of fluorine on the crystallization and properties of glass-ceramic materials of the SiO_2-CaO-MgO-Al_2O_3 system. Due to its good mechanical properties, we intend to obtain Diopside-type pyroxene as the main crystalline phase in the synthesized glass-ceramics. We are interested in the addition of fluorine due to its potential dual role as a flux and as a nucleating agent during crystallization of the studied glass compositions.

EXPERIMENTAL DETAILS

Two parent glass compositions are studied, which correspond to the (60-X)SiO_2-23CaO-12MgO-5Al_2O_3-XCaF_2 chemical formula (with compound proportions in mol.%), and in which two CaF_2 levels (X = 3 and X = 6 mol.%) are investigated. The base composition for these parent glasses, which correspond to X = 0, is formulated within the primary crystallization field of Diopside in the quaternary system SiO_2-CaO-MgO-Al_2O_3. Two ~ 200g-batch compositions are prepared by mixing together (for 1h in a ball mill) stoichiometric amounts of silica sand (purity of 99.98 wt.%, average particle size of 110μm, MERASI, México); calcite ($CaCO_3$, purity of 99.6 wt.%, average particle size of 45μm, Materiales La Gloria, México); dolomite ((Ca,Mg)$(CO_3)_2$, with ~3 wt.% of $CaCO_3$ as main impurity, average particle size of 35μm, Química del Rey, México); Al_2O_3 (HPA-0.5, purity of 99.99 wt.%, average particle size of 0.75μm, Sasol, EUA), and fluorite (CaF_2, with 10.3 wt.% SiO_2, 2.2 wt.% Al_2O_3 and 19.8 wt.% $CaCO_3$ as main impurities, average particle size of 30μm, Materiales La Gloria, México). Then, the batch mixtures are melted at 1450°C for 3h in a Pt crucible using a Lindberg/Blue M BF51433 PC-1 high temperature electric furnace. Glasses in bulk form are produced by casting of the melts on a preheated stainless steel plate, which is immediately followed by annealing at 600°C for the next 5h and then by cooling at a rate of 1°C/min until room temperature is reached. Glass-ceramic materials are obtained by subjecting the bulk glasses to a thermal cycle comprising the nucleation and growth stages depicted in Figure 1. The morphology and size of the Diopside crystals are analyzed with the help of Scanning Electron Microscopy (SEM), using a Phillips XL30 ESEM, with an accelerating voltage of 20 KV, backscattered electron imaging mode and working distance of 10 mm. Prior to this, the glass-ceramic samples are mounted in cold-cure epoxy resin and then ground, polished and graphite-coated using standard ceramographic techniques.

X-Ray Diffraction (XRD) is used to verify the amorphous state of the parent glasses, as well as to identify the formed crystalline phases in the glass-ceramic materials and to estimate the degree of crystallinity of the latter. This is carried out employing a Philips X'Pert 3040 diffractometer, CuKα radiation, accelerating voltage of 40 kV and current of 30 mA, in the 2θ range of 10-80°, with a step size of 0.035° (2θ/s). The degree of crystallinity of the glass-ceramics, X(%), is estimated by using a method developed by S.M. Ohlberg and D.W. Strickler [7], which employs equation (1):

$$X(\%) = \frac{I_g - I_x}{I_g} \times 100 \tag{1}$$

where I_g and I_x are the integrated intensities of the background lines of the XRD patterns obtained for a parent glass and for the corresponding glass-ceramic material, respectively.

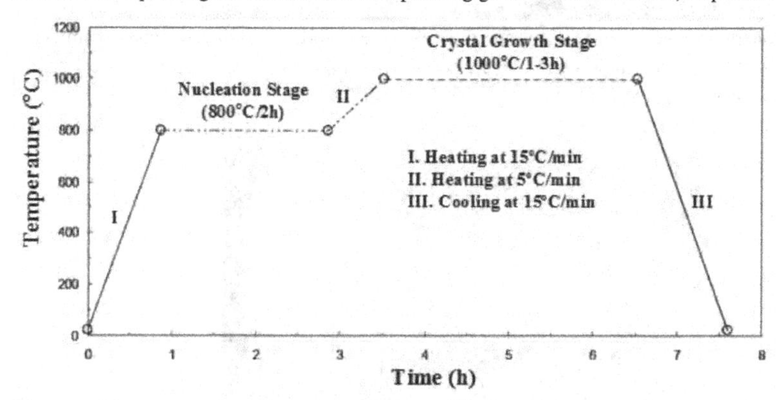

Figure 1. Heat treatment cycle used for the crystallization of the parent glass monoliths.

The apparent density of the glass-ceramics is determined by the Archimedes' principle in distilled water at room temperature. The Vickers microhardness (HV) of the glass-ceramic materials is measured using a Wilson Tukon 300-FM apparatus, applying a load of 500gf; ten indentations are made per each sample; their mean value is reported in this work.

DISCUSSION

Both parent glass compositions are transparent and colorless. The bulk glasses undergo surface crystallization, with the formed crystals growing towards the core of the bulk glass as crystallization proceeds. The samples are fully crystallized by using a holding time of 3h for the crystal growth stage of the employed heat treatment cycle. See Figures 2 and 3. According to the XRD patterns and SEM micrographs shown in Figure 4, the glass-ceramic materials obtained for the two studied compositions show the formation of acicular crystals of a Diopside-type pyroxene phase, which are homogeneously distributed in the glassy matrix. The specimen with the least amount of added fluorine shows the highest microhardness value, as well as the largest and thickest acicular crystals of Diopside-type pyroxene, the lowest apparent density and the largest crystallized fraction. The values obtained by us for these properties are shown in Table I.

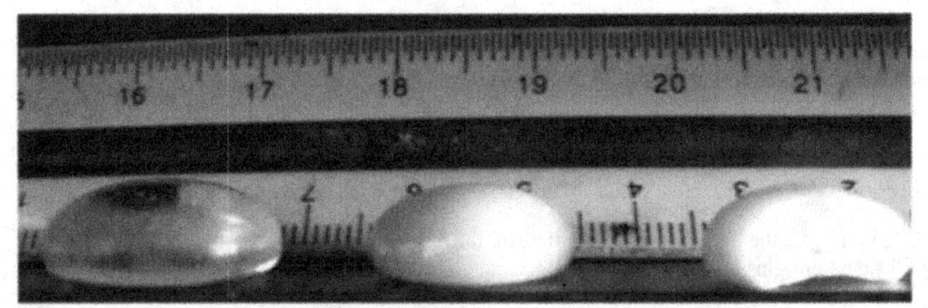

Figure 2. Transparent and colorless parent glass (left) and fully crystallized glass-ceramics (center and right) obtained with a thermal cycle comprising a holding time of 3h for the crystal growth stage. Upper scale is in cm.

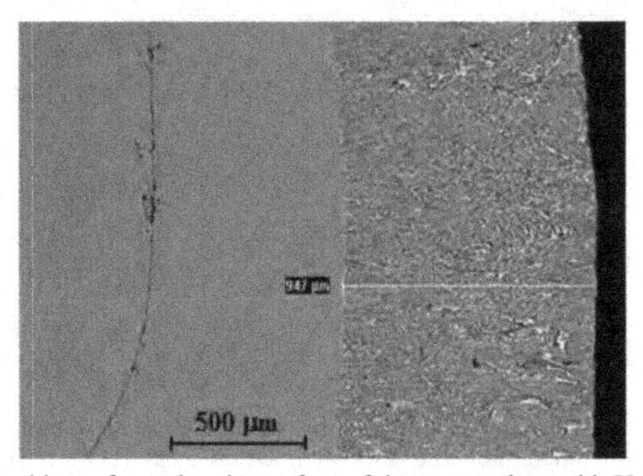

Figure 3. Crystallized layer formed at the surface of the parent glass with X = 3, after a thermal cycle comprising a holding time of 1h for the crystal growth stage.

Since the glasses developed in this investigation are prone to surface crystallization, our results indicate that CaF_2 added in the amounts used by us does not act as nucleating agent. However, it does affect the growth of the acicular crystals of the Diopside-type pyroxene phase. The length, thickness and amount of these crystals, as evidenced by the diminution observed in the crystallized fraction given in Table I and the microstructures shown in Figure 4, tend to decrease with increasing addition of CaF_2. This could be related to the fact that the addition of fluorine disrupts the glass structure by producing non-bridging oxygen ions. This lowers the viscosity of the glass and facilitates the motion, and homogenization, throughout the entire glass network of ions such as Ca^{2+}, Mg^{2+} and Al^{3+}. These ions, in the absence of a significant amount of fluorine, must take part in the formation of the Diopside-type pyroxene phase by creation of localized regions enriched with them. Thus, an increased addition of CaF_2 delays the formation of the expected crystalline phase. This results also in a diminution of the microhardness of the glass-ceramic materials. On the other hand, the small increment observed in the apparent density of the glass-ceramics with increasing fluorine addition could be related to the replacement in the materials' chemical composition of a small amount of a compound having a slightly smaller density (SiO_2, which has a theoretical density of 2.8 g/cm^3 for quartz) by a small amount of another one having a slightly higher density (CaF_2, which has a theoretical density of 3.18 g/cm^3) [8].

Lastly, based on our results, we propose that the materials developed in this study may be considered as viable alternatives for applications in abrasive and corrosive environments, as well as for substrates for metallic coatings, and for abrasion-resistant floor tiles and other structural applications.

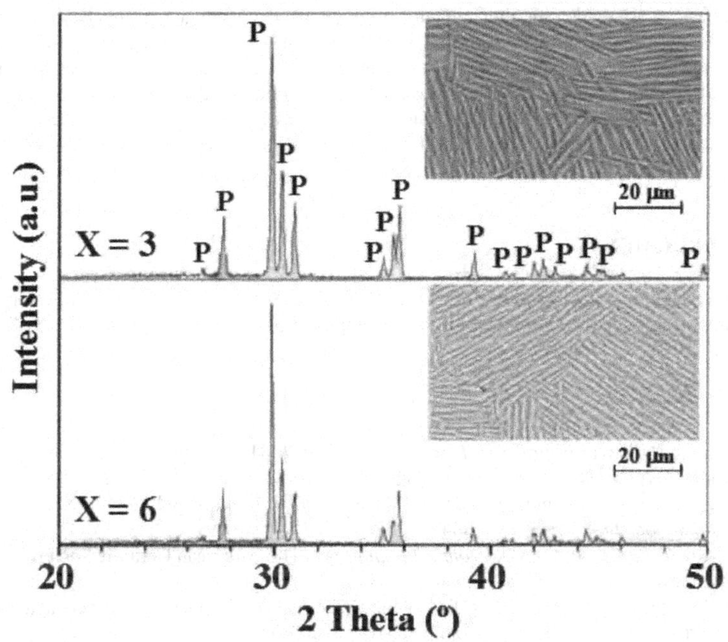

Figure 4. XRD patterns and SEM micrographs corresponding to fully crystallized glass-ceramic samples obtained for the two studied compositions. Key: P = Diopside-type pyroxene phase.

Table I. Apparent density, Vickers microhardness (HV), crystallized fraction and crystal size determined for the synthesized glass-ceramics.

Level of CaF_2 addition (X, mol.%)	Apparent density (g/cm^3)	HV (Kg$_f$/cm^2)	Crystallized fraction (%)	Crystal size (µm)
3	3.18	698	43.5	~10-40
6	3.20	685	42.8	~10-30

CONCLUSIONS

The materials developed in this study may be considered as viable alternatives for applications in abrasive and corrosive environments, as well as for substrates for metallic coatings, and for abrasion-resistant floor tiles and other structural applications. Although the bulk glasses are prone to surface crystallization, with the formed crystals growing towards the core of the bulk glass as crystallization proceeds, the samples could be fully crystallized by using a sufficiently long holding time for the crystal growth stage of the heat treatment cycle. The XRD and SEM analyses carried out confirm the formation of acicular crystals of Diopside-type pyroxene phase. CaF_2 added in the amounts used by us does not act as nucleating agent. However, it does affect the growth of the acicular crystals of the Diopside-type pyroxene phase. This in turn affects some important properties of the synthesized glass-ceramics. The specimen with the least amount of added fluorine shows the highest microhardness value, as well as the largest and thickest acicular crystals of Diopside-type pyroxene, the lowest apparent density and the largest crystallized fraction. This is attributed mainly to the effect of fluorine on the glass structure and properties.

ACKNOWLEDGMENTS

The authors express their gratitude to CONACYT, U.A. de C. and Cinvestav-Saltillo for the financial support and facilities provided for the development of this work.

REFERENCES

1. J.Mª. Fernández Navarro, *Bol. Soc. Esp. Ceram. Vidrio* **7** [4], 431-458 (1968).
2. Z. Strnad, *Glass Ceramic Materials*, (Elsevier, Praga, 1986).
3. D.U. Tulyaganov, S. Agathopoulos, H. R. Fernandez, J.M. Ventura and J.M.F. Ferreira, *J. Eur. Ceram. Soc.* **24**, 3521-3528 (2004).
4. W. Höland and G. Beall, *Glass-Ceramic Technology*, (The American Ceramic Society, Westerville, Ohio, USA, (2002).
5. D.U. Tulyaganov, S. Agathopoulos, J.M. Ventura, M.A. Karakassides, O. Fabrichnaya and J.M.F. Ferreira, *J. Eur. Ceram. Soc.* **26**, 1463-1471 (2006).
6. A. Rafferty, A. Clifford and R. Hill, *J. Am. Ceram. Soc.* **83** [11], 2833-2838 (2000).
7. S.M. Ohlberg and D.W. Strickler, *J. Am. Ceram. Soc.* **45** [4], 170-171 (1962).
8. A. Al-Noaman, S.C.F. Rawlinson and R.G. Hill, *J. Non-Cryst. Solids* **358**, 1850-1858 (2012).

Mater. Res. Soc. Symp. Proc. Vol. 1485 © 2013 Materials Research Society
DOI: 10.1557/opl.2013.213

Effect of Controlled Corrosion Attack With HCl Acid on the Fatigue Endurance of Aluminum Alloy AISI 6063-T5, under Rotating Bending Fatigue Tests

G. M. Domínguez-Almaraz[1,2], J. L. Ávila-Ambriz[1], F. Peyraut[2], E. Cadenas-Calderón[1]

[1]Facultad de Ingeniería Mecánica, Universidad Michoacana de San Nicolás de Hidalgo (UMSNH), Santiago Tapia No. 403, Col. Centro, Morelia, Michoacán 58000, México.
[2]LaboratoireM3M-IRTES (EA 7274), Université de Technologie de Belfort-Montbéliard, 90010 Belfort, France

ABSTRACT

Corrosion attack is implemented on the aluminum alloy AISI 6063-T5 for six different non corroded and pre-corroded specimens. Concerning pre-corroded specimens, they are divided in two groups; the first one is immersed for 1 and 2 minutes in hydrochloric acid with 20% concentration, and the second group for 2, 4, 6 minutes of immersion but in HCl with 38% of concentration. Rotating bending fatigue tests are carried out on corroded and non-corroded specimens at the frequency of 50 Hz, at room temperature and without control of environmental humidity. Loading conditions are fixed by Finite Element numerical simulation; the loading ranges are 90%, 80%, 70% and 60% of the yield stress of this aluminum alloy. A numerical simulation study is carried out by means of the Ansys software to investigate the stress concentration factor variation induced by the proximity of two close pitting holes: in longitudinal and transversal direction regarding the principal applying loading. Finally, optical microscopy is used to analyze the fracture surfaces in longitudinal and transversal directions, in order to establish possible causes of fatigue fracture.

Keywords: Corrosion attack, AISI 6063-T5, stress concentration factors, corrosion pits.

INTRODUCTION

Industrial applications of 6063-T5 aluminum alloy involve very good corrosion resistance with good weldability features, high machinability and adequate fatigue strength. It presents suitability for anodizing and enables decorative usage in architectural applications like sections for windows, doors and curtain walls. It is also used in truck and trailer flooring, pneumatic installation, irrigation pipes, ladders and railings[1-4]. The influence of corrosion on the fatigue performance of metallic components is of considerable importance for industrial applications and the case of 6063-T5 aluminum alloy is not the exception; the present study has been devoted to assess the pre-corrosion effect on rotating bending fatigue endurance of this material. Corrosion pit size distributions and fatigue lives have been studied extensively over the past 20 years [5]; furthermore, the application of aluminum alloys has been wide expanded in car, aeronautical, and other industries in the last years [6]. Nowadays, the knowledge of the mechanical-corrosion behavior of aluminum alloys is of principal interest for industrial use [7].

EXPERIMENTAL DETAIL

The rotating bending fatigue machine to carry out this study is described elsewhere [8-9]. In the present study rotating bending fatigue tests are carried out with hourglass shape specimens at a frequency of 50 Hz, without control of environmental humidity and surface roughness. Nevertheless, all specimens are machined similarly in order to maintain the surface roughness with no large variation before the pre-corroding process. Fig. 1a shows the sample dimensions. Chemical composition in weigh and principal mechanical properties of testing material are shown in Table 1a and b respectively. The pre-corrosion process is controlled by immersion of specimens in two acid solutions with different concentration. The first one is a hydrochloric acid solution at concentration of 38% and pH close to 0.8; in this case the time of immersion is 2,4 and 6 minutes. The second solution has a 20% concentration of HCl and a pH close to 0.42 at room temperature; in this case the exposed corrosion time is 1 and 2 minutes.

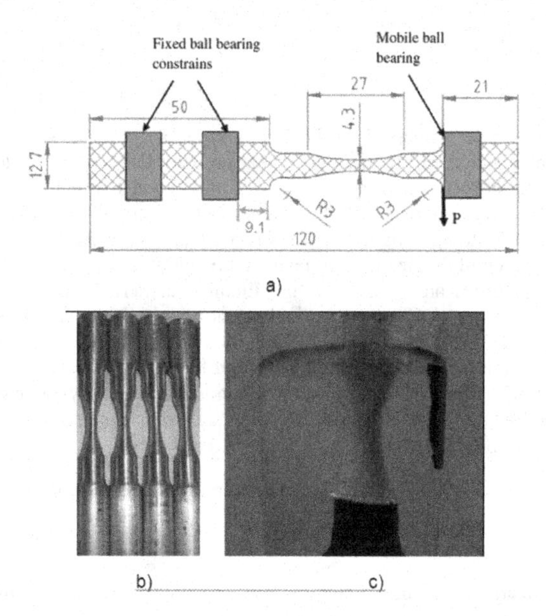

Figure 1. (a) Dimensions of specimen, (b) Non-corroded specimens, (c) corrosion attack process.

Table 1. a) Chemical composition in weigh percentages of aluminum alloy 6063-T5.

Si	Fe	Cu	Mn	Mg	Cr	Zn	Ti	Al
0.3-0.6	0.35	0.1	0.1	0.4-0.85	0.1	0.1	0.1	>96.9

Table 1. b) Principal mechanical properties of aluminum alloy 6063-T5.

Elastic Limit MPa	UTS MPa	% elongation	Young Modulus	Hardness Brinell
145	187	0.33	68.9	60

Figure 2 shows the experimental results. The pre-corroding process induces an evident decrease on fatigue endurance for this aluminum alloy. The specimens without corroding pre-treatment show higher fatigue lives than the pre-corroded specimens for all values of applied load σ_a: 60%, 70%, 80% and 90% regarding the yield stress of this material.

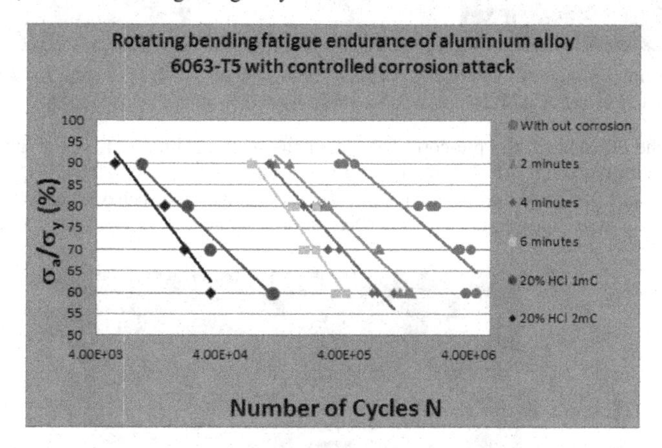

Figure 2. Tendency lines and fatigue life for 6 types of aluminum specimens, subjected to rotating bending fatigue tests.

As shown in Figure 2, the higher the pre-corrosion attack time the shorter is the fatigue life. Additionally this difference becomes higher for the low loading regimes (60% and 70% of yield stress). The specimens attacked with a lower concentration of HCl, which present the lower pH, have lower fatigue endurance.

DISCUSSION

The average size of pitting holes on corroded specimens at 38% of HCl ranges between 150 mm and 250 μm after 2 and 4 min of corrosion, respectively. Pitting holes density on specimen surface increases with time of corrosion attack rather than increasing the pitting holes size [9]. This tendency is seen in Figure 3, the longest time of corrosion attack induces coalescence of neighboring pitting holes forming "grooves" along the longitudinal direction of specimen.

As it has been demonstrated by different authors [11-12] pitting corrosion is initiated frequently at the grain boundaries. Since the extrusion process implies an elongation of grain boundaries parallel to the specimen longitudinal direction, this should be at the origin of the coalescence of pitting holes and formation of grooves e.g., see Figure 3c. Here, the proximity of two contiguous corrosion pitting holes is related to the stress concentrations factors, accelerating the fracture initiation and crack propagation; as a consequence, the fatigue endurance of the AISI 6063-T5 aluminum alloy decreases dramatically (Figure 2).

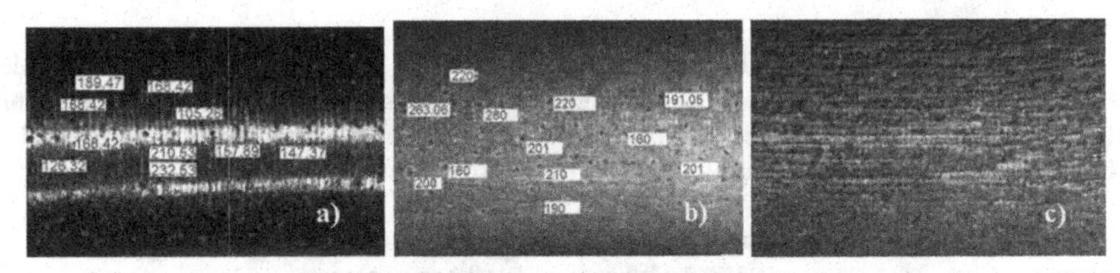

Figure 3. Pitting diameter of corroded specimens before testing with (a) 2 min, (b) 4 min and (c) 6 min of corrosion attack with HCl at a concentration of 38% and pH close to 0.8.

On the other hand, the pre-corroded specimens with a solution of HCl at 20%, seems to have superficial corrosion effect in the case of one minute of corrosion attack, Fig. 4a; whereas for the 2 minutes pre-corroded specimens, Fig. 4b, there is a similar behavior as to that observed for 6 minutes of acid exposure at 38% of concentration, Fig. 3c.

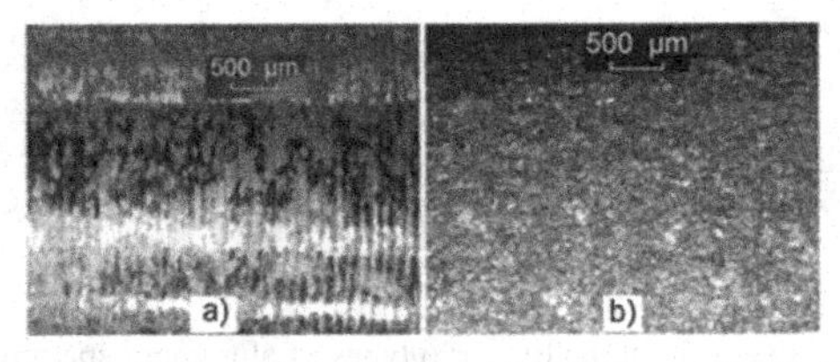

Figure 4. Corroded specimens before testing after (a) 1 min, and (b) 2 min in HCl at concentration of 20% with pH close to 0.42.

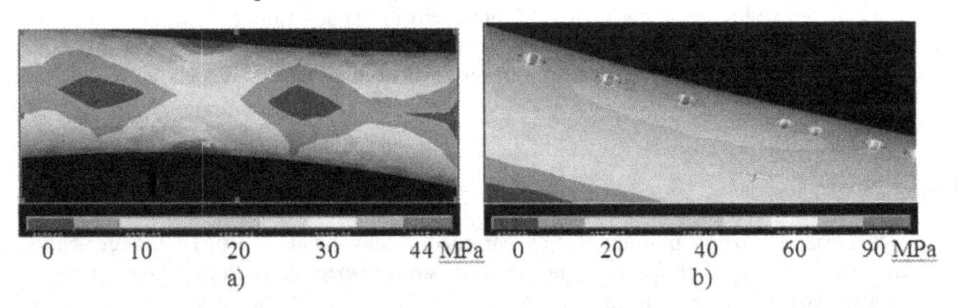

Figure 5. (a) Von Mises stresses distribution for non pre-corroded specimen, σ_{max}=44 MPa, (b) Von Mises stress on pre-corroded specimen, σ_{max}= 90 MPa. Both tests with P = 1.45 Kg, Fig. 1a.

The observed reduction on fatigue endurance can be associated with pitting holes. Crack initiation has been identified with the presence of one, two or more neighboring corrosion pitting holes. In order to model the pitting corrosion effect on the stress concentration factor K_t, a finite element analysis is undertaken as shown in Figure 5.

For the same loading conditions, the maximum Von Mises Stress increases 2.04 times for the pre-corroded specimen as compared to a clean virgin sample, see Fig.5 a-b. Additionally, it is interesting to analyze the interaction between two neighboring pitting holes both in a parallel and perpendicular direction with respect to the loading axis. Numerical simulation shows that stress concentration factor K_t increases dramatically as shown in Figure 5. This is particularly clear for the perpendicular direction to the loading stress. All specimens have been identically loaded and those without pitting reach a maximum Von Mises stress of 44 MPa. However, two pitting holes in longitudinal direction give rise to a maximum stress of 90 MPa with 50 μm of separation (Figure 6a), and 95 MPa for 5 μm of separation (Figure 6c). As for the perpendicular direction with respect to the loading axis, the maximum Von Mises stress becomes 104 MPa with 50 μm of separation (Figure 6b), and 368 MPa with 5 μm of separation (Figure 6d).

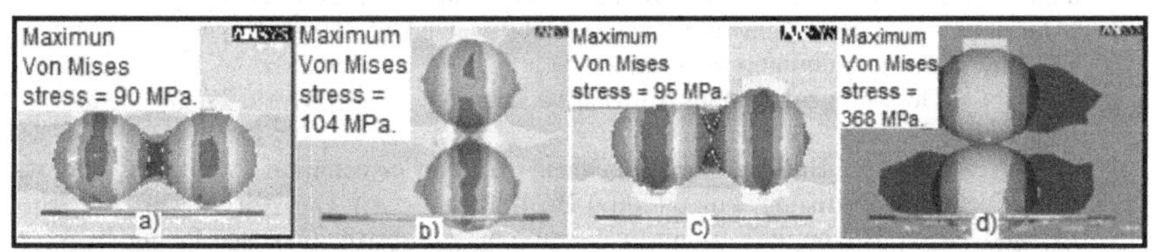

Figure 6. Interaction between 2 contiguous pitting holes separated 50μm a) and b), versus 5 μm of separation c) and d).

Analysis on fracture surfaces shows that crack initiates at corrosion pitting holes and propagates by following a path of neighboring pits (transversal pits), Fig. 7.

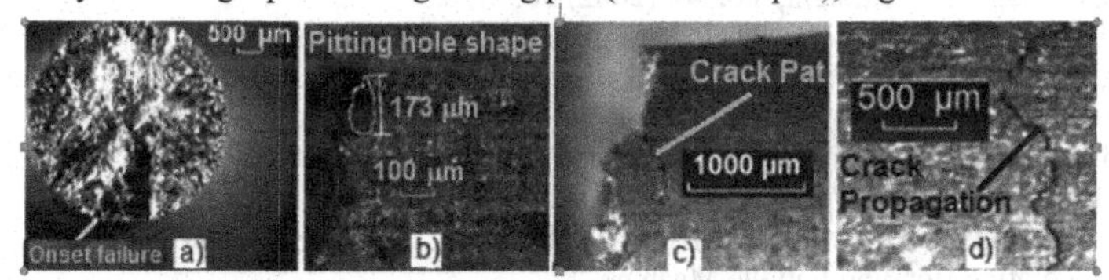

Figure 7. (a) Frontal fracture image shows the beginning of failure, (b) lateral fracture image at the same point to onset failure, (c), crack path and (d) crack propagation.

CONCLUSIONS

- Rotating bending fatigue endurance decreases with time of pre-corrosion attack. Fatigue endurance separation between pre-corroded specimen increases when the applying load decreases. The last traducing the fact that at high loading, pre-corrosion effect is reduced.
- The corrosion potential increases when the acid concentrations decrease (lower pH).
- The stress concentrations zones are related to corrosion pitting holes in this study. Furthermore, stress concentration factor increase dramatically with the pitting separation, particularly for the perpendicular disposition of pitting with regard of loading axis.

- The stress concentration factor Kt is higher in reducing the distance between the pitting holes; this should lead to an exponential law between these two parameters. Crack paths following always the transversal direction as shown in experimental results, Fig. 7.

ACKNOWLEDGMENTS

We express our gratitude to the University of Michoacan (UMSNH) in Mexico and the UTBM in France for the facilities received to carry out this study. A special mention of gratitude to CONACYT (The National Counsel for Science and Technology in Mexico) for the financial support destined to this research work.

REFERENCES

1. J. Hirsch, "Virtual Fabrication of Aluminum Products (Microstructural Modeling in Industrial Aluminium Fabrication Process)", Wiley-VCH, First Ed. (2006).
2. "Aluminum for Future Generations: Sustainability Update 2005" (Haymarket, London, UK: International Aluminum Institute, 2005).
3. S. K. Das, "Designing Aluminum Alloys for a Recycle-Friendly World," *Light Metal Age*, June (2006) 26–33.
4. G.M. Scamans, N. Birbilis, R.G. Buchheit, "3.08 - Corrosion of Aluminum and its Alloys", Shreir's Corrosion, Vol. 3 (2009) 1974-2010.
5. J.J. Medved, M. Breton, P.E. Irving, "Corrosion pit size distributions and fatigue lives—a study of the EIFS technique for fatigue design in the presence of corrosion", *International Journal of Fatigue,* Vol. 26 (2004) 71–80.
6. I.N. Fridlyander, V.G. Sister, O.E. Grushko, V.V. Berstenev, L.M. Sheveleva and L.A. Ivanova, "Aluminum Alloys: Promising Materials in the Automotive Industry", *Metal Science and Heat Treatment,* Vol. 44 (2002) 365-370.
7. C.J. Villalobos-Gutiérrez, G.E. Gedler-Chacón, J.G. La Barbera-Sosa, A. Piñeiro, M.H. Staia, J. Lesage, D. Chicot, G. Mesmacque and E.S. Puchi-Cabrera, "Fatigue and Corrosion Fatigue Behavior of an AA6063-T6 Aluminum Alloy Coated with a WC–10Co–4Cr Alloy Deposited by HVOF Thermal Spraying", *Surf. & Coating Tech. 2012,* Vol. 18 (2008) 4572-4577.
8. G. M. Dominguez Almaraz, M. Guzmán Tapia, Patent No. 276818, Mexico (2010); "High Speed Rotating Bending Fatigue Machine".
9. G. M. Domínguez Almaraz, J. L. Ávila Ambriz, E. Cadenas Calderón and J. J. Villalón López (2012), "Rotating Bending Fatigue Endurance and Effect of Controlled Corrosion on the AISI-SAE 6063-T5 Aluminum Alloy", MRS Proceedings/Vol. 1373/2012, doi:10.1557/opl.2012.308.
10. H. Sahlaoui, K. Makhlouf, H. Sidhom, J. Philibert, "Effects of aging conditions on the precipitates evolution, chromium depletion and intergranular corrosion susceptibility of AISI 316L: experimental and modeling results", *Mat. Sci. and Eng.,* Vol. 372 (2004), 98-118.
11. H. Kokawa, M. Shimada, Z.J. Wang, Y.S. Sato, M. Michiuchi, "Grain boundary engineering for intergranular corrosion resistant austenitic stainless steel" *Key Eng. Mat,* Vol. 261-263 (2004), 1005-1010.

Mater. Res. Soc. Symp. Proc. Vol. 1485 © 2013 Materials Research Society
DOI: 10.1557/opl.2013.214

The Influence of Modifiers on the Pigmentary Properties of Titanium Dioxide

Marta A. Gleń[1] and Barbara U. Grzmil[1]
[1]West Pomeranian University of Technology, Szczecin, Institute of Chemical and Environment Engineering, 70-322 Szczecin, Pułaskiego 10, E-mail: mglen@zut.edu.pl

ABSTRACT

In the present work modified titanium dioxide products are prepared. The influence of introduced modifiers (phosphorus, potassium, aluminium, and tin) on the photoactivity, optical properties, and phase composition of titanium dioxide is studied. The molar contents of P_2O_5, K_2O, Al_2O_3, and SnO_2 in relation to TiO_2 are 0.10, 0.18, 0.24, and 0.08-1.32 mol%, respectively. The research is aimed at obtaining the pigmentary rutile TiO_2 with the highest possible photostability and improved optical properties.

INTRODUCTION

Titanium dioxide is a very important and multifunctional material. TiO_2 possesses various interesting properties: optical, catalytic and dielectric, which lead to many industrial applications [1]. The main use of titanium dioxide is as a white pigment [2]. The manufacturing process of TiO_2 pigment is aimed at obtaining the product with the lowest photocatalytic activity [3]. There are several methods which lead to the decrease of TiO_2 photoactivity, therefore stability and fastness of materials containing pigment can be improved. On the one hand, it can be realized through selection of binder type, the addition of radical scavengers and/or UV absorbers and the use of UV absorbing clear coats [4]. The other methods include modification of titanium dioxide during pigment processing.

One of the most effective methods decreasing photoactivity of titanium dioxide is introduction of defects into the crystal lattice through selective ion doping [5,6] in order to generate electron traps for the holes formed as a result of radiation. The photocatalytic activity of titanium dioxide has been demonstrated to be a complex function of dopant concentration, distribution of dopants, their energy level within titanium dioxide lattice, electronic configurations, electron donor concentration, and light intensity [7]. The optical response of material is also determined by its chemical composition, atomic arrangement and physical dimension [8]. Unfortunately, many different modifiers which strongly absorb UV light and decrease titanium dioxide photoactivity cause undesired colouration of titanium dioxide white pigments [9-12].

The other way to increase the photostability of titanium dioxide is retarding radical formation at the pigment surface. This method comprises removal of oxygen, water and hydroxyl groups from titanium dioxide surface and pigment microencapsulation [13]. The latter method consists of covering the pigment surface with a layer which prevents the electron-hole pair from redox cycle on the pigment surface [4,13].

In the light of the presented information it has been found that the defecting of titanium dioxide atomic structure in order to obtain photostable pigmentary titanium dioxide with suitable optical properties is a technique which requires a large amount of work, especially experimental and scientific. In the present work modifiers such as P_2O_5, K_2O, Al_2O_3, and SnO_2 are introduced into hydrated titanium dioxide. The research is aimed at obtaining the pigmentary rutile TiO_2 with the lowest possible photoactivity.

EXPERIMENT

The starting material is concentrated suspension of technical-grade hydrated titanium dioxide (HTD). The modifiers solutions (calculated to P_2O_5, K_2O, Al_2O_3, and SnO_2) are introduced to. HTD. The molar contents of P_2O_5, K_2O, and Al_2O_3 in relation to TiO_2 are constant and are 0.10, 0.18, and 0.24 mol%, respectively. Molar contents of SnO_2 in relation to TiO_2 are 0.08–1.32 mol%. The contents of introduced modifiers, P_2O_5, K_2O, Al_2O_3, and SnO_2, are verified by ICP-AES analysis (Optima 5300 DV, Perkin–Elmer). The obtained pulp after thorough mixing (mechanical stirrer, 25 rpm, time 0.5 h) is heated in laboratory muffle furnace (LM 312.13) to an assumed temperature. The prepared samples are calcined with gradually increasing process temperature corresponding to the conditions of the commercial calcination. Final calcination temperature, at which anatase-rutile phase transformation rate is >97%, depended on the kind and amount of modifiers introduced to HTD.

The crystallinity of heat-treated titanium dioxide products is examined by powder X-ray diffraction (X'Pert PRO Philips diffractometer, CuK_α radiation). Relative amounts of anatase and rutile phase are calculated from the diffraction intensities corresponding to (101) reflection of anatase and (110) reflection of rutile. The mass fraction of rutile in titanium dioxide powders, W_R, is determined from the following equation:

$$W_R = 100/(1+I_A/k \cdot I_R) \qquad (1)$$

where: I_A and I_R are the peak intensities of anatase (101) and rutile (110) and k is the coefficient (the ratio of peak intensity (101) 100 % of anatase to the peak intensity (110) 100 % of rutile).

The optical properties of rutile titanium dioxide powders are characterized by measuring the colour in the white system and in the grey system. In the white system, brightness and white tone and in the grey system, relative lightening power (Tinctorial Strength - *TcS*) and grey tone (Spectral Characteristics - *SCx*) of titanium dioxide products are determined. The white paste is composed of reference or examined titanium dioxide, linseed-tung oil and colloidal silica. The grey paste consisted of reference or examined titanium dioxide, linseed-tung oil, colloidal silica and carbon black. The procedure of the test using Konica Minolta CM-600d Spectrophotometer, (Standard Illuminant C, 2° Standard Observer) is described elsewhere [14]. The optical properties are calculated from X, Y and Z values using CIE XYZ system where Y correlates with the lightness and two mutually orthogonal axes X and Z are indirectly associated with the tone.
The relative lightening power (*TcS*) and grey tone (*SCx*) are calculated from the following equations:

$$TcS_s = TcS_r + (Y_s - Y_r)\,100 \qquad (2)$$

$$SCx_s = SCx_r + (Z_s - X_s) - (Z_r - X_r) \qquad (3)$$

where: TcS_s, SCx_s – the relative lightening power and grey tone of the examined titanium dioxide, TcS_r, SCx_r – known values of the relative lightening power and grey tone of the reference titanium dioxide, X_s, Y_s, Z_s – average values of trichromatic components X, Y, Z of the

examined titanium dioxide, X_r, Y_r, Z_r – known average values of trichromatic components X, Y, Z of the reference titanium dioxide.
The brightness and white tone are calculated from the equations:

$$B_s = B_r + (Y_s - Y_r) \quad (4)$$

$$WT_s = WT_r + (Z_s - X_s) - (Z_r - X_r) \quad (5)$$

where B_s, WT_s – the brightness and white tone of the examined titanium dioxide, B_r, WT_r – known values of the brightness and white tone of the reference titanium dioxide.

The photoactivity of the prepared titanium dioxide products is characterized by the white lead-glycerine test. The procedure of the test involved the preparation of an aqueous paste containing a titanium dioxide product, glycerine, colloidal silica and basic lead carbonate. The obtained paste is illuminated with UV-Vis light for 1 h with a radiation intensity of 500 W/m^2 (climatic chamber SUNTEST XLS+, Atlas, Xenon lamp 290 – 800 nm). During the exposure of titanium dioxide to the UV-Vis radiation the photooxidation of glycerine is followed by reduction of Pb^{2+} to Pb^0. The discolouration of the paste induced by photoreaction is evaluated by measuring the ΔE^* parameter using CIE $L^*a^*b^*$ system (Konica Minolta CM-600d Spectrophotometer, Standard Illuminant C, 2° Standard Observer). The lower ΔE^*, the more photostable pigment is obtained. The colour stability (photoactivity) is calculated from the measurements of $L^*a^*b^*$ values of titanium dioxide products before and after UV-Vis irradiation from the equation:

$$\Delta E^* = \sqrt{\Delta L^{*2} + \Delta a^{*2} + \Delta b^{*2}} \quad (6)$$

where ΔL^* - the lightness change of titanium dioxide product after irradiation with determined dose, Δa^*, Δb^* - the colour change of titanium dioxide product after irradiation with determined dose.

DISCUSSION

In the following experiments pigmentary rutile titanium dioxide modified with P_2O_5, K_2O, Al_2O_3, and SnO_2 is obtained. The influence of modifiers on the anatase-rutile phase transformation is followed by X-ray diffraction analysis. On the basis of obtained XRD patterns, degrees of transformation are calculated according to Equation 1. The results of final calcination temperature (T_F) at which anatase-rutile transformation rate is >97% are shown in Figure 1. Unmodified titanium dioxide containing 98.5 mass% of rutile is obtained at 890 °C. An addition of phosphorus, potassium, and aluminium caused a decrease in rutilization degree. Thus, the final calcination temperature (T_F) of modified titanium dioxide is increased to 1050°C. Moreover, the final calcination temperature of modified titanium dioxide products with rutilization degree >97% is increased when SnO_2 is introduced additionally. An increasing amount of this modifier inhibited anatase-rutile phase transformation in the presence of P_2O_5, K_2O, and Al_2O_3.

The optical properties and photocatalytic activity of titanium dioxide modified with the increasing content of SnO_2 and constant amount of other modifiers (P_2O_5, K_2O, Al_2O_3) are determined. The objective of these studies is to observe the influence of modifiers on the colour and photostability of titanium dioxide products assigned for pigmentary applications.

The optical properties of unmodified and modified titanium dioxide are characterized by the colour measurements in the white system and in the grey system. The optical properties of titanium dioxide samples determined in the white system are brightness (Figure 2) and white tone (Figure 3), whereas relative lightening power (Figure 4) and grey tone (Figure 5) are measured in the grey system. The brightness (*B*), white tone (*WT*), relative lightening power (*TcS*), and grey tone (*SCx*) of unmodified titanium dioxide are 94.33, -9.03, 1310, and -0.43, respectively. The same properties for TiO_2-PKAl are 93.44, -9.04, 1560, and -0.46, respectively. It is observed that with the increasing content of SnO_2 in rutile TiO_2, modified with phosphorus, potassium and aluminium, the relative lightening power and grey tone are the most improved for TiO_2-PKAlSn0.64. The changes of brightness and white tone values are insignificant.

It is found that the changes of photocatalytic activity are insignificant when increasing contents of tin in modified TiO_2 are introduced.

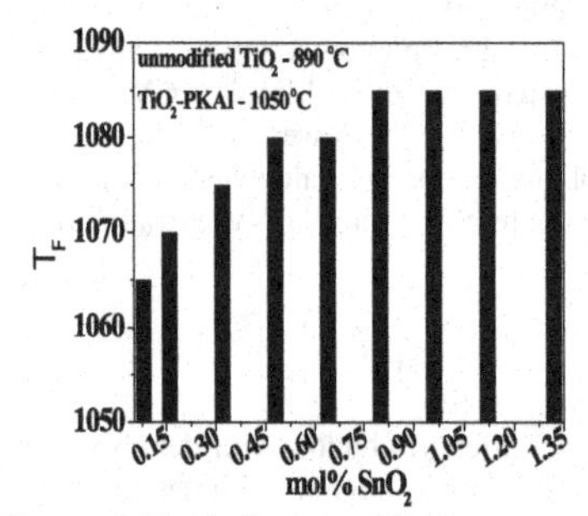

Figure 1. The influence of SnO_2 contents (in the composition of modifiers: P_2O_5, K_2O, Al_2O_3) on the final calcination temperature (T_F) of titanium dioxide.

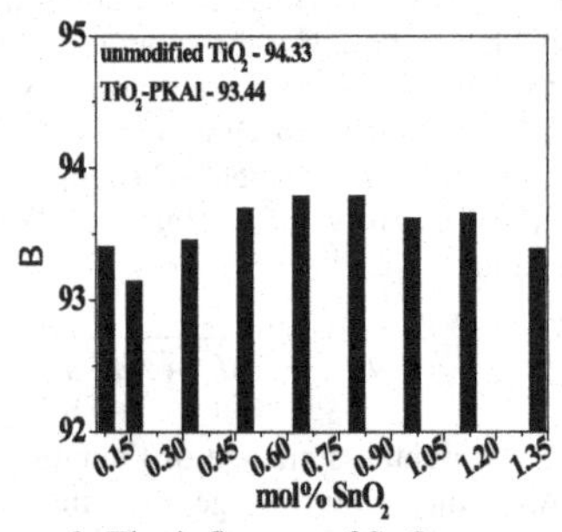

Figure 2. The influence of SnO_2 contents (in the composition of modifiers: P_2O_5, K_2O, Al_2O_3) on the brightness (B) of titanium dioxide.

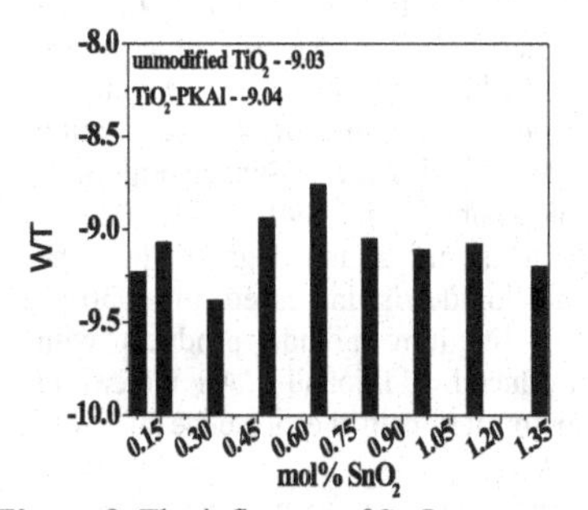

Figure 3. The influence of SnO_2 contents (in the composition of modifiers: P_2O_5, K_2O, Al_2O_3) on the white tone (WT) of titanium

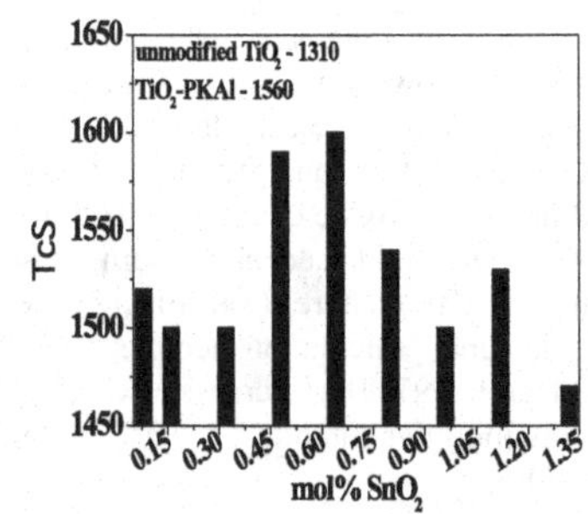

Figure 4. The influence of SnO_2 contents (in the composition of modifiers: P_2O_5, K_2O, Al_2O_3) on the relative lightening power (TcS)

dioxide.

of titanium dioxide.

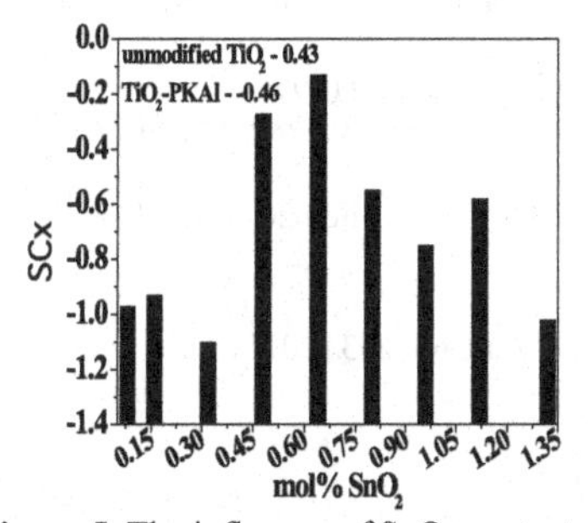

Figure 5. The influence of SnO_2 contents (in the composition of modifiers: P_2O_5, K_2O, Al_2O_3) on the grey tone (SCx) of titanium dioxide.

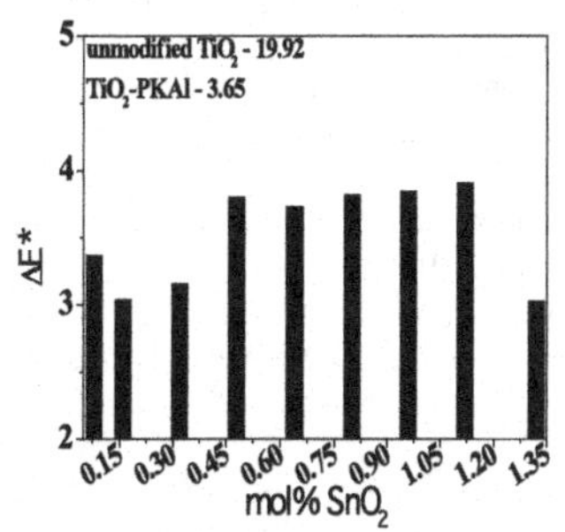

Figure 6. The influence of SnO_2 contents (in the composition of modifiers: P_2O_5, K_2O, Al_2O_3) on the photoactivity (ΔE^*) of titanium dioxide.

CONCLUSIONS

Increasing amounts of SnO_2 inhibit the anatase-rutile phase transformation in the presence of P_2O_5, K_2O, and Al_2O_3.

The changes of brightness, white tone values and photoactivity are insignificant when increasing contents of tin in modified TiO_2 are introduced.

It is observed that with the increasing contents of tin in rutile TiO_2, modified with phosphorus, potassium and aluminium, the relative lightening power and grey tone is the most improved for TiO_2-PKAlSn0.64 .

The promising direction of investigations over these interesting modifiers, leading to obtain the TiO_2 with improved optical properties and high photostability, can be the studies of these additives at different amounts, different way of calcination or different phase composition with other compounds.

ACKNOWLEDGMENTS

This work is funded from financial support on science for 2008–2011.

REFERENCES

1. M. Inagaki, Y. Nakazawa, M. Hirano, Y. Kobayashi and M. Toyoda, *Int. J. Inorg. Mater.* **3**, 809 (2001).
2. S. Karvinen, *Solid State Sci.* **5**, 811 (2003).
3. A. Mills and S. Le Hunte, *J. Photochem. Photobiol. A: Chem.* **108**, 1 (1997).
4. M. P. Diebold, *Surf. Coat. Int.* **6**, 250 (1995).
5. W. Choi, A. Termin and M. R. Hoffmann, *J. Phys. Chem.* **98**, 13669 (1994).
6. A. W. Xu, Y. Gao and H. Q. Liu, *J. Catal.* **207**, 151 (2002).

7. J. C. Yu, J. Lin and R. W. M. Kwok, *J. Phys. Chem. B* **102**, 5094 (1998).
8. X. Chen and S. S. Mao, *Chem. Rev.* **107**, 2891 (2007).
9. A. Di Paola, E. Garcia-Lopez, S. Ikeda, G. Marci, B. Ohtani and L. Palmisano,*Catal. Today* **75**, 87 (2002).
10. K. Wilke and H. D. Breuer, *J. Photochem. Photobiol. A: Chem.* **121**, 49 (1999).
11. J. Rademachers and H. P. Heisse, US Patent 4 917 735, Filed 18 January 1989, Issued 17 April 1990
12. H. Knittel, R. Bauer, E. Liedek and G. Etzrodt. US Patent 4 844 741, Filed 20 April 1988, Issued 4 July 1989
13. F. K. Tyler, *Paint Coat. Ind.* **16**, 32 (2000).
14. M. Gleń, B. Grzmil, J. Sreńscek-Nazzal and B. Kic, *Chem. Pap.* **65**, 203 (2011).

Mater. Res. Soc. Symp. Proc. Vol. 1485 © 2013 Materials Research Society
DOI: 10.1557/opl.2013.215

Phase Formation at Selective Laser Synthesis in Al_2O_3–TiO_2-Y_2O_3 Powder Compositions

M. Vlasova[1], M. Kakazey[1], P. A. Márquez-Aguilar[1], A. Ragulya[2], V. Stetsenko[2], and A. Bykov[2]

[1]Center of Investigation in Engineering and Applied Sciences of the Autonomous University of the State of Morelos (CIICAp-UAEMor), Av. Universidad, 1001, Cuernavaca, Mexico.

[2]Institute for Problems of Materials Science, National Academy of Sciences of Ukraine, 3, Krzhyzhanovsky St., Kiev, 252680, Ukraine

ABSTRACT

The phase formation in the zone of directional laser irradiation of compacted Al_2O_3–TiO_2–Y_2O_3 mixtures has been investigated. It is established that phase formation is carried out within the framework of binary mixtures Al_2O_3 – Y_2O_3 and Y_2O_3 – TiO_2.

INTRODUCTION

Laser synthesis of ceramics from powder mixtures is a perspective technological method of obtaining new materials with complex properties which are formed in conditions of directional high-speed and high-temperature heating and subsequent cooling with high-speed. Feature of this synthesis is the phase formation in a narrow area of radiation and is determined by the regime of radiation, and also by heat and temperature-conductivity material and a number of other technical factors [1-3]. The complexity, variety and transience of the processes occurring in a moving zone of irradiation require research of the phase formation in such metastable conditions. This problem is the most actual when the result of irradiations is the formation of products of interaction with the elements of surface texture. The aim of this work is investigation of phase formation in zone of laser treatment of ternary (Al_2O_3–TiO_2–Y_2O_3) powder mixtures.

EXPERIMENTAL PROCEDURE

In the present work, specimens are prepared from analytically pure Al_2O_3, TiO_2, and Y_2O_3 powders (produced by REASOL). Powder mixtures x wt. % Al_2O_3–y wt. % TiO_2 – z wt. % Y_2O_3 are compacted in pellets with a diameter of 18 mm and a thickness of 2–3 mm under a pressure of 300 MPa. The compositions of compacts are presented in Table 1. The compositions of the mixtures are calculated so that, with increased in the Al_2O_3 content in the mixture, the TiO_2/Y_2O_3 molar ratio will be 0.25.

Laser treatment is performed in an LTN-103 unit (continuous-action laser with $\lambda = 1064$ nm). The power of radiation (P) is 120 W, the diameter of the beam (d) is 1.5 mm, and the linear traversing speed of the beam is $v = 0.15$mm/s.

The synthesis products are investigated by the X-ray diffraction (XRD) method in Cu K_{α} radiation (a DRON-3M installation, Russia). An electron microscopy study is performed with a HU-200F type scanning electron microscope and a LEO 1450 VP unit.

RESULTS

As a result of irradiation, concave channels (tracks) are formed on the surfaces of compacted specimens (figure 1 a). Their formation is associated with development of high temperatures in the irradiation zone, due to which the melting–solidification processes proceed simultaneously [2, 4]. These tracks are easily removed from the compacts. The numbers of tracks correspond to the numbers of used mixtures. The concave ceramic tracks consist from the $Y_3Al_5O_{12}$ (or YAG), $Y_2Ti_2O_7$ and Al_2O_3 (see table 1, figure 2). Predominant phase is the$Y_3Al_5O_{12}$. With increasing Al_2O_3 content ($C_{init.}$) in initial mixtures up to 70 mol.% the content of all newly formed phases increases. At the same time at $C_{init.} > 70$ mol.% the content of $Y_3Al_5O_{12}$ and $Y_2Ti_2O_7$ decreases, but content of corundum continues to increase (figure 3).

The deviation of the peak intensities (that is amplitudes) of the lines of all phases from standard ratios [5] indicates on theirs texturing. In tables II- IV, the values of the texture coefficients (TC) for different phases in tracks Nos. 1/1–1/7 are presented. The calculation is performed by the formula [6]:

$$TC(hkl) = \frac{I(hkl)}{I_o(hkl)}\left\{\frac{1}{n}\sum\frac{I(hkl)}{I_o(hkl)}\right\}^{-1},$$

where I(hkl) are the measured intensities of the (hkl) reflection, I_o(hkl) are the intensities according to the JCPDS cards, and n is the number of reflections used in the calculation.

As follows from tables II – IV, all crystalline phases carry the signs of texturing. The most noticeable texturing of Y3Al5O12 and Y2Ti2O7 occurs when in tracks are a minimum content of corundum (at Cinit.< 60 mol.%). From data of tables II-IV, it is possible conclude that in various places of track the texturing of Y3Al5O12 and Y2Ti2O7 crystals carried out in different directions. This can be seen in figures 1, b- d. On cross-section of track it is seen (figure 1, d) that the principal axes of the dendrites are oriented from a cold zone (zone of sintering) toward the surface. That is, the texturing is realized not only in the direction of movement of the laser beam (see figures 1, a, b), but also in the direction perpendicular to the plane of the track. At the same time, a rapid cooling of surface of tracks leads to rapid crystallization and formation of crystallites (or dendrites) of small sizes in different directions with different TC (see figure 1, a).

Thus, we can conclude that at directed laser beam motion a rapid heating of the powder material in the horizontal and vertical directions with a certain temperature gradients is realized. As a consequence, in pellets there are different temperature conditions of interaction between the components of mixtures, formation of new phases and their crystallization.

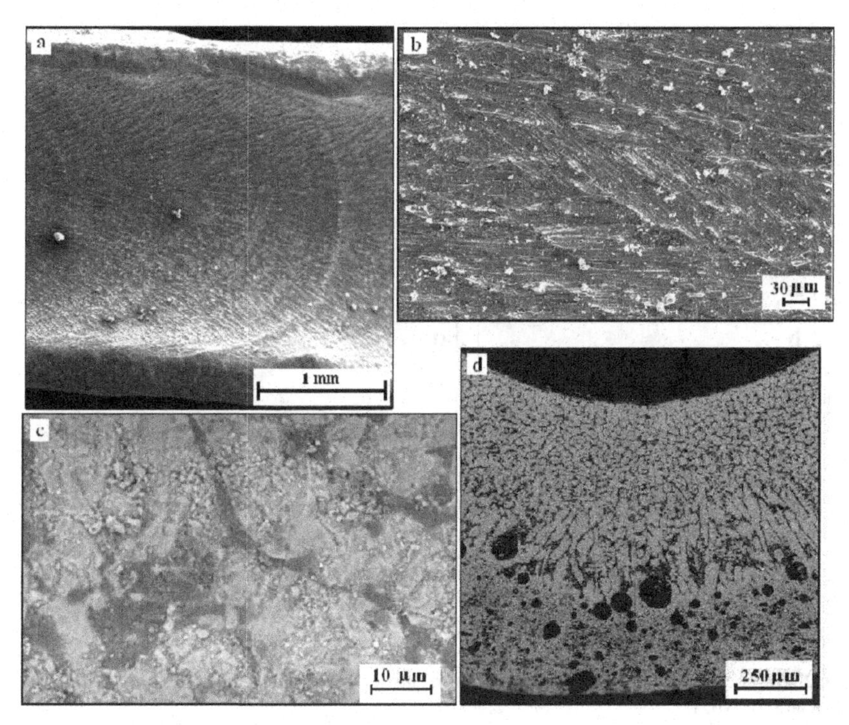

Figure 1. Electron micrographs of track No.1/4. The top view of the track (a, b); longitudinal section (c); the cross section of track (d).

Table I. Phase composition in initial mixtures and in zone of tracks.

Number of specimen	Composition of powder mixture			Phase composition of track
	Al_2O_3 (mol.%)	Y_2O_3 (mol.%)	TiO_2 (mol.%)	
1/1	50	40	10	$Y_3Al_5O_{12}$, little $Y_2Ti_2O_7$, traces Y_2TiO_5
1/2	55	36	9	$Y_3Al_5O_{12}$, little $Y_2Ti_2O_7$
1/3	60	32	8	$Y_3Al_5O_{12}$, little $Y_2Ti_2O_7$, traces Al_2O_3
1/4	65	28	7	$Y_3Al_5O_{12}$, little Al_2O_3, little $Y_2Ti_2O_7$
1/5	70	24	6	$Y_3Al_5O_{12}$, Al_2O_3, little $Y_2Ti_2O_7$
1/6	75	20	5	$Y_3Al_5O_{12}$, Al_2O_3, little $Y_2Ti_2O_7$
1/7	70	16	4	$Y_3Al_5O_{12}$, Al_2O_3, little $Y_2Ti_2O_7$

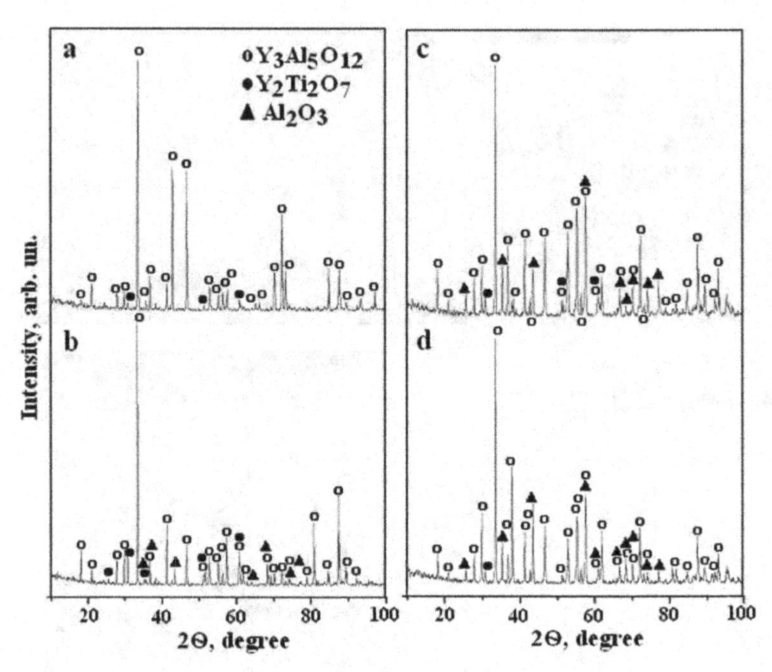

Figure 2. Fragments of X-ray diffraction patterns of tracks. (a) in tracks No. 1/2, (b) No. 1/4, (c) No. 1/6, and (d) No. 1/7.

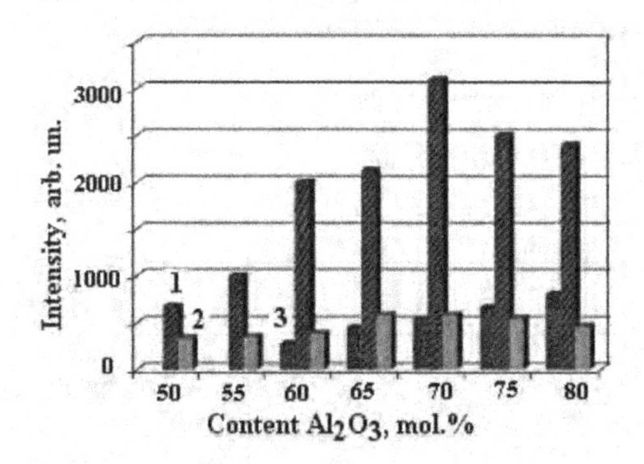

Figure 3. Change of the intensity of diffraction lines (squares under the respective peaks) depending on content of Al_2O_3 in initial mixtures. (1) for the $Y_3Al_5O_{12}$ with d = 0.268 nm; (2) for the $Y_2Ti_2O_7$ with d = 0.291 nm; (3) for the Al_2O_3 with d = 0.255 nm

Table II. Values of the coefficients of texture (TC) for Y3Al5O12 in the tracks.

Phase	N track	C_{Al2O3}, mol.%	TC for <hkl>									
			211	321	400	420	422	521	532	444	640	842
$Y_3Al_5O_{12}$	1/1	50	0.34	0.39	0.29	**2.18**	0.27	0.76	0.65	0.72	0.96	**4.5**
	1/2	55	0.07	0.34	0.3	1.1	0.73	0.57	**2.7**	0.66	0.28	**3.3**
	1/3	60	0.53	0.8	0.53	1.18	0.72	**1.62**	0.82	0.68	1.24	**1.75**
	1/ 4	65	0.87	1.04	0.9	**1.52**	0.67	**1.77**	1.06	1.08	0.52	0.52
	1/6	75	0.3	**1.4**	1.22	1.05	0.58	1.05	1.08	1.16	1.09	**1.86**
	1/7	80	047	0.72	0.72	0.72	1.31	**1.43**	0.08	0.023	1.63	**1.42**

Table III. Values of the coefficients of texture for Y2Ti2O7 in the tracks.

Phase	N track	C_{Al2O3}, mol.%	TC for <hkl>				
			222	400	331	440	622
$Y_2Ti_2O_7$	1/1	50	0.59	0.59	0.95	0.13	**2.7**
	1/2	55	0.13	0.83	1.03	**2.26**	0.77
	1/3	60	0.06	0.06	**3.6**	0.37	0.63

Table 4.Values of the coefficients of texture for Al2O3 in the tracks.

Phase	N track	C_{Al2O3}, mol.%	TC for <hkl>						
			012	104	110	113	214	300	1010
Al_2O_3	1/3	60	0.11	0.48	**3.25**	0.19	0.42	0.63	1.94
	1/ 4	65	0.17	0.12	0.17	0.41	0.23	0.11	**5.8**
	1/6	75	0.49	0.78	0.51	0.76	0.13	0.27	**4.0**
	1/7	80	0.13	0.35	0.35	0.72	**4.4**	0.27	0.64

CONCLUSIONS

Investigations have shown that at the laser irradiation of mixtures of x mol. % Al_2O_3 – y mol. % TiO_2 – z mol. % Y_2O_3 at TiO_2/Y_2O_3 = 0.25:

• phase formation is carried out within the framework of binary mixtures Al_2O_3 – Y_2O_3 and Y2O3 – TiO2 and accompanied by the formation of Y3Al5O12, Y2Ti2O7an excess ofAl2O3;

• the texture and microstructure of formed crystalline body depends on the ratio of the phases formed during rapid cooling of the melt.

ACKNOWLEDGMENTS

The authors wish to thank CONACYT for financial support (Project 155731).

REFERENCES

1.V. Shishkovskii, *Laser synthesis of functional mesostructures and 3D parts* (Moscow, Fizmatlit, 2009).

2. M.Vlasova, M.Kakazey, P. A. Márquez Aguilar in *Advances in ceramics. Synthesis and characterization, processing and specific application,* edited by Costas Sikalidis (INTECH, Open Access Publisher, 2011) pp.393-420.

3. V. M. Orera, R. I. Merino, J. A. Pardo, A. A. Larrea, J. I. Peña, C. Gonzalez, P. Poza, J. Y. Pastor, J. LLorca, *Acta Mater*, **48** (18/19), 4683 (2000).

4. M. Vlasova, M. Kakazey, B. Sosa Coeto, P. A. Marquez Aguilar, I. Rosales, A. Escobar Martinez, V. Stetsenko, A. Bykov, A. Ragulya, *Sci.Sintering*, **44**, No.2, 17 (2012).

5. *JCPDS*, Swarthmore, PA (1996).

6. S. Ruppi, *Int. J. Refract.Met.Hard Mater*. **23**, Iss. 4–6, 306 (2005).

Mater. Res. Soc. Symp. Proc. Vol. 1485 © 2013 Materials Research Society
DOI: 10.1557/opl.2013.251

Nanostructured Ceramic Oxides Containing Ferrite Nanoparticles and Produced by Mechanical Milling.

A. Huerta-Ricardo[1], K. Tsuchiya[2], T. Umemoto[2] and H. A. Calderon[1]
[1]Departamento de Física, ESFM-IPN, UPALM Ed. 9, Zacatenco D.F. Mexico.
[2]Dept. Prod. Systems Eng., Toyohashi University of Technology, Toyohashi Aichi 441 Japan.

ABSTRACT

This investigation deals with the production process and the characterization of ceramic materials consisting of magnetic particles in an insulating matrix. Composites made of magnetite particles (Fe_3O_4 or $MgFe_2O_4$) in a wüstite or magnesiowüstite matrix (Fe_xO or $Mg_{1-x}Fe_xO$), respectively, have been produced by means of mechanical milling and spark plasma sintering. As-milled powders have a nanocrystalline structure in both systems. As a function of milling time, low energy milling gives rise to an increasingly higher volume fraction of wüstite in the Fe_xO-Fe_3O_4 system while it promotes increasing amounts of magnesiowüstite ($Mg_xFe_{1-x}O$). Sintering is performed from 673 to 1273 K in vacuum. Sintering at low temperatures allows retention of nanosized grains containing a fine dispersion of magnetic particles in a wüstite and magnesiowüstite matrix. Measurement of magnetic properties reflects the constitution of the sintered samples and the effect of grain size. It also allows determination of the transformation sequence both during mechanical milling and sintering

INTRODUCTION

Oxide ceramics are under intense investigation for their technological advantages in magnetization, dielectric response and chemical stability in such diverse applications as magnetic recording media, induction cores and microwave resonant circuits [1, 2]. Several researchers have mixed iron oxides with other ceramics as MgO, NiO, and ZnO to improve mechanical and magnetic properties [2, 3]. In such systems, the matrix (NaCl-type structure) has paramagnetic behavior and ferromagnetism is provided by precipitation of a second phase (spinel structure). Groves and Fine have shown that precipitation of a coherent spinel-structure in a NaCl-structure matrix may occur merely by rearrangement of the concentration of cations and the interstitial–site occupancy [4]. This transformation gives an octahedral precipitate morphology. However, clear evidence regarding the phase transformation mechanism in oxide ceramics is not well known. These materials are normally produced by conventional techniques, for example by diffusing iron into the matrix and thermal treatment to induce precipitation.

The present investigation deals with the production of magnetite precipitates (Fe_3O_4) in a wüstite matrix (Fe_xO) and magnesioferrite precipitates ($MgFe_2O_4$) in a magnesia matrix (MgO). The matrix phases have a NaCl structure and the precipitates a cubic inverse spinel structure. In such systems, the phase decomposition process occurs by the rearrangement of the cations in the continuous oxygen lattice. The combination of a nonconductive matrix and magnetic particles is interesting due to the expected magnetic and physical properties. The spatial and size distribution of precipitates can affect the nanocomposite properties. In addition, the expected effect of the nanosize grains needs to be documented in order to determine the relationship between nanostructure and magnetic and physical properties in oxide ceramics.

EXPERIMENTAL PROCEDURE

Both systems have been produced by the mechanochemical reaction of pure components in the form of powders of Fe_3O_4 and C; Fe_3O_4 and Fe; Fe_2O_3 and Fe in one case and MgO and Fe_2O_3 in the other system. Powders have been commercially acquired with a high purity level. All powder handling has been performed in an Ar-filled glove box with an oxygen content below 0.01 ppm. Low, high and ultrahigh energy mills are used (horizontal, planetary and Spex mills). Table I gives the chemical composition of the prepared samples. The ball milling process has been periodically interrupted in order to monitor the microstructural development of the mixtures. Consolidation of powders is performed by spark plasma sintering (SPS). Cylindrical sintered samples of 18 mm or 13 mm in diameter and 5 mm of thickness are produced by sintering at high temperature (873 to 1273 K) and low pressure (50 MPa) and at low temperature (673, 773 K) and high pressure (100 MPa).

Table I. Materials and chemical compositions.

ID	Composition	ID	Composition
Fe_3O_4	100% Fe_3O_4	Pure MgO	MgO
Fe_3O_4+C	Fe_3O_4-3 wt.% C	0.5 cat.% Fe	MgO-1 wt.% Fe_2O_3
Fe_3O_4+Fe (I)	Fe_3O_4-15 wt.% Fe	1 cat.% Fe	MgO-2 wt.% Fe_2O_3
Fe_3O_4+Fe (II)	Fe_3O_4-12.5 wt.% Fe	4 cat.% Fe	MgO-8 wt.% Fe_2O_3
Fe_2O_3	100 % Fe_2O_3	6 cat.% Fe	MgO-11 wt.% Fe_2O_3
Fe_2O_3+Fe	Fe_2O_3-15 wt.% Fe	10 cat.% Fe	MgO-18 wt.% Fe_2O_3
		20 cat.% Fe	MgO-33 wt.% Fe_2O_3
		40 cat.% Fe	MgO-57 wt.% Fe_2O_3

Characterization is done by means of X-Ray diffraction (Siemens D-500, CuKα, XRD), scanning electron microscopy (JSM-200, SEM)), and transmission electron microscopy (Hitachi 800, TEM). Measurement of magnetic properties is made in a vibrating sample magnetometer (Riken Denshi Co., VSM). Identification of the iron containing phases is done by Mössbauer spectroscopy (Source Co-60, MöS).

RESULTS AND DISCUSSION

Figure 1 shows a representative sequence of X ray diffraction patterns taken as a function of milling time for the wüstite-magnetite system and the sample Fe_3O_4+Fe (I). Milling in this case is done in a horizontal mill (low energy) and thus the relatively long processing times. The angular positions for the pure phases are also indicated in this figure. The patterns show that only broad diffraction peaks develop as milling takes place. Thus a grain size reduction can be expected i.e. the development of a massive nanostructure as milling proceeds. There is also a variation of the unit cell dimensions since the experimental peak angular positions disagree to either of the equilibrium phases (Fe, Fe_xO or Fe_3O_4). This can be interpreted either as a Fe-lean non-equilibrium wüstite phase or as a Fe-rich non-equilibrium magnetite. However, the direction of the observed displacements suggests that Fe-lean wüstite is formed during milling. Increasingly longer milling periods produce XRD patterns with better match to the expected pattern for wüstite. Thus the XRD results in Fig. 1 indicate the formation of a single phase after a relatively long milling time in this low energy milling process.

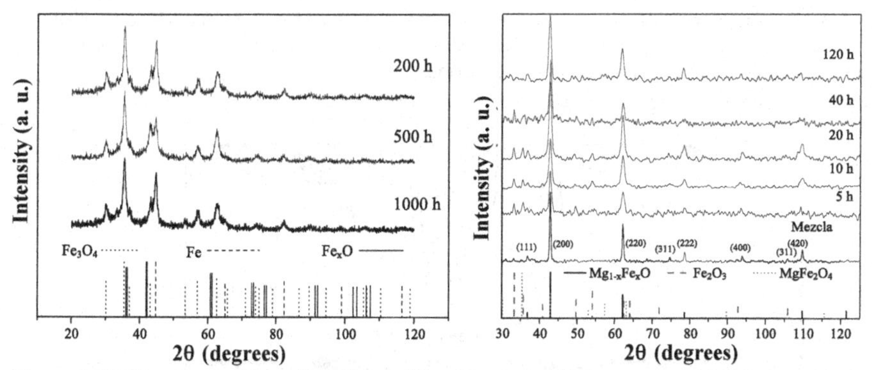

Figure 1. XRD patterns of powders after milling (horizontal mill) for the sample Fe_3O_4+Fe (I), see Table I. The milling time is indicated.

Figure 2. XRD patterns of MgO-4 cat.% Fe before and after milling in a Spex mill.

The MgO-$MgFe_2O_4$ system has been prepared in two different mills, a planetary (high energy) and a Spex (ultrahigh energy) mill. XRD patterns show that low amounts of Fe_2O_3 can be dissolved into the MgO lattice producing distortions to the MgO lattice. Samples with higher Fe_2O_3 contents show some reflections corresponding to this compound but also displaced MgO and $MgFe_2O_4$ peaks (see Fig. 2). According to these results, the MgO lattice saturates with Fe and thus the as-milled powders are composed of $Mg_{1-x}Fe_xO$, $MgFe_2O_4$ and possibly retained Fe_2O_3. Relatively long milling periods in this mill produce a reduction of the crystal size as indicated by the broadening of the diffraction maxima.

The use of Mossbauer spectroscopy allows a detailed characterization in some of these materials. The Mössbauer spectrum (MS) of a sample can be formed by three or more spectra at a time, depending on the different Fe occupied crystallographic sites in the sample. For example for the as-milled powders in Fig. 2, the Mössbauer spectra in Fig. 3 elucidate the phases in the massive nanostructure after 40 and 120 h of milling. The MS in Fig. 3a (40 h of milling) shows the overlapping of two sextets (hyperfine or Zeeman interaction, magnetic effect) and two duplets (quadrupole interaction, paramagnetic effect). Magnesiowüstite can be related to the duplets due to its structure (NaCl type) and magnetic behavior. The duplets are related to the isomeric unfolding of the Fe^{3+} ion (high spin) and the exchange of two different positions between the Fe^{2+} and Fe^{3+} ions (low spin), respectively. Magnesioferrite has a spinel structure that gives rise to two sextets in the MS. One of them comes as a result of Fe ions in tetrahedral positions and the other one from Fe ions in octahedral positions. Quantitatively the amount of magnesioferrite (75 %) is much higher than that of magnesiowüstite (around 25 %). Interestingly the MS of Fig. 3b (120 h of milling) shows additional features since there are two duplets (I), a monolet or isomer shift (II) and a sextet (III). The duplets can be related to magnesiowüstite as before and the sextet is due most likely to Fe that should be also responsible for the isomer shit or single shift. This last feature shows the presence of Fe^{3+} ions in the paramagnetic phase. Quantitatively the relative amount of magnesioferrita is reduced to 20.5 % while that of magnesiowüstite increases to around 76 %. Thus according to these results, the mechanochemical reaction between powders ends before 120 h of milling, the $MgFe_2O_4$ phase forms at relatively short milling times (lower than 40 h) and that it transforms to magnesiowüstite if milling proceeds to longer periods.

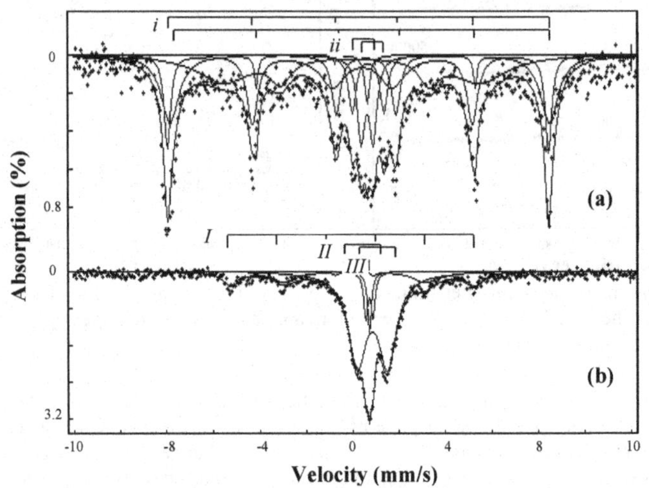

Figure 3. Mossbauer spectra from sample MgO-4 % cat. Fe after milling (a) 40 h and (b) 120 h in a Spex mill. Details in text.

Figure 4 presents measurements of magnetization (M) as a function of applied field (H) in the wüstite-magnetite system after mechanical milling in a horizontal mill. Fig. 4a depicts magnetization curves for powders milled 50 to 1000 h of the Fe_3O_4+Fe (I) sample. The saturation magnetization (M_s) at 15 KÖe decreases as a function of milling time. This behavior can be explained by the transformation of magnetic phases into non-magnetic phases i.e., magnetite and/or iron into wüstite. The original powders in Fe_3O_4-Fe (I) are magnetic and the formation of wüstita by milling gives rise to a reduction of the saturation magnetization as shown in Fig. 4b. A net reduction of approximately 50% is achieved when comparing M_s values after 50 and 1000 h of milling. The central sections of the hysteresis loops are magnified and shown as an insert of Fig. 4a. This can be used to evaluate the coercivity of the as-milled powders (half width of hysteresis loop with no induced field, M). As seen in Fig. 4a, the coercivity remains constant in the as-milled powders of the Fe+Fe_3O_4 (I) sample as a function of the milling times, indicating that the size of the magnetic domains is independent of the milling time. On the other hand, Fig. 4b shows a summary of results (M_s Vs milling time) for the different samples investigated. There is a tendency for M_s to decrease as a function of milling time in samples having magnetic phases (Fe_3O_4 and/or Fe) as starting materials. Clearly the formation of the paramagnetic phase Fe_xO can explain such a trend. However, the transformation of magnetic phase into paramagnetic wüstite is not complete after 1000 h of milling due to the applied low energy processing. A different behavior is seen for the sample prepared with a Fe_2O_3. There is an increase of the M_s value for milling times up to 500 h. Additional milling promotes M_s values similar to those found for other samples. Apparently Fe_2O_3 transforms into Fe_3O_4 upon milling and then a transformation to Fe_xO takes place. Therefore the decomposition sequence appears to be $Fe+Fe_2O_3 \rightarrow Fe_3O_4 \rightarrow Fe_xO$. This can also be predicted by the relative values of the oxides free energy of formation [5]. Thus the initial components have a strong influence on the development of phases.

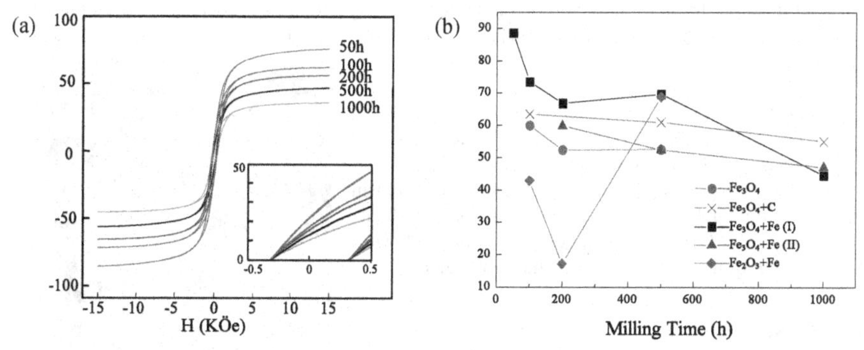

Figure 4. Magnetization curves of as-milled powders in $FeO-Fe_3O_4$ system. (a) $Fe+Fe_3O_4$ (I) sample. (b) Saturation magnetization (M_s) as a function of milling time.

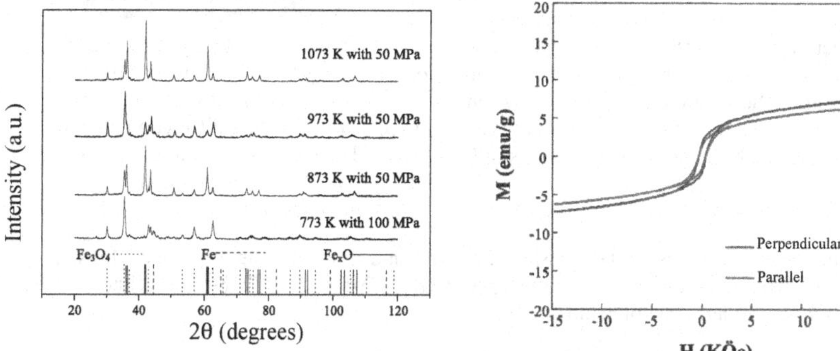

Figure 5. XRD patters of sintered Fe_3O_4 + Fe (I) sample at different temperatures.

Figure 6. Magnetization curves for Fe_3O_4 + Fe (I) sample after milling 500 h and sintering at 900 °C.

Sintered composite samples present phases that correspond to the processing temperature. Low sintering temperatures below 600 °C promote formation of a nanostructure with a similar distribution of phases as in the as-milled powders. Higher processing temperature renders mostly stable wüstite as a matrix with dispersed magnetite in rather small grain sizes. Figure 5 shows XRD patterns for sample Fe_3O_4 + Fe (I) after sintering from 500 to 800 °C. Analysis of the patterns shows the development of wüstita with relatively higher amounts towards higher temperatures. There is a rather slight peak sharpening suggesting a rather fine gain size. Lower sintering temperatures retain the as milled phases with larger volume fractions of non equilibrium magnetite in this sample. Fig. 6 shows a representative example of the magnetic response of this set of samples. Magnetic measurements in solid samples have been performed both parallel and perpendicularly to the sintering load axis on specimens cut along such directions. There is a small difference in the absolute value of the induced magnetization that indicates a preferential alignment of the magnetic domains in the radial direction of the sintered samples. As can be seen by comparing results in Figs. 4 and 5, the values of saturation are somewhat different for this sample before and after sintering with a light decrease in the sintered specimens. This confirms the XRD results regarding the formation of stable wüstita and a dispersion of magnetite.

Figure 7 shows a TEM dark field image of sample Fe_3O_4 + Fe (I) after 1000 h of milling and sintering at 1000 °C. The insert shows the corresponding diffraction pattern with satellites characteristic of a spinodally decomposing system. The modulated variation of composition is indicated by an arrow. The formation of a composite material with a fine dispersion of magnetic centers is thus confirmed.

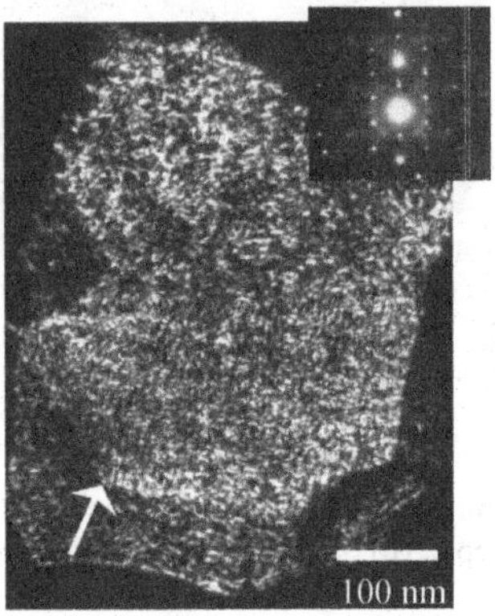

CONCLUSIONS

A nanostructured composite of magnetic domains in a paramagnetic matrix is demonstrated in two systems i.e., Fe_3O_4 – Fe_XO and $Mg_XFe_{(1-X)}O$-$MgFe_2O_4$. Phase development agrees with the relative values of the energy of formation. The spinel phase precedes the formation of the oxygen solid solution in both systems under investigation. Different milling conditions have been used but processing can be controlled to achieve similar results.

Figure 7. TEM dark field image of Fe_3O_4 + Fe (I) by using a (002) reflection. Insert shows the corresponding diffraction pattern. The periodic variation of contrast reflects the composition modulations.

ACKNOWLEDGMENTS

The use of facilities at Toyohashi University and a scholarship from the Ministry of Education of Japan are gratefully acknowledged. CONACYT and ICYT-DF are acknowledged for financial support and a scholarship.

REFERENCES

1. I. Yamaguchi, T. Manabe, T. Kumagai, W. Kondo and S. Mizuta. *J. Mater. Res.* **13** (1998) 935.

2. G. W. Groves and M. E. Fine, "Solid Solution and Precipitation Hardening in Mg-Fe-O Alloys", *J. Appl. Phys.* **35** (1964) 3587.

3. R. Pandey, J. D. Gale, S.K. Sampath and J. M. Recio, "Atomistic Simulation Study of Spinel Oxides: Zinc Aluminate and Zinc Gallate" *J. Am. Ceram. Soc.* **82** (1999) 3337.

4. R. C. O'Handley. *Modern Magnetic Materials*. John Wiley &Sons, Inc. (1999) 432.

5. C.H.P. Lupis, *Chemical Thermodynamics of Materials*, Prentice-Hall Inc. A Simon and Schuster Company, Englewood Cliffs, New Jersey, (1983).

Mater. Res. Soc. Symp. Proc. Vol. 1485 © 2013 Materials Research Society
DOI: 10.1557/opl.2013.252

Recycled HDPE-tetrapack composites. Isothermal crystallization, light scattering and mechanical properties

A Parada-Soria[1], HF Yao[1], B Alvarado-Tenorio[1], L Sanchez-Cadena[2] and A Romo-Uribe[1,*]
[1] Laboratorio de Nanopolimeros y Coloides, Instituto de Ciencias Fisicas, Universidad Nacional Autonoma de Mexico, Cuernavaca Mor. 62210, MEXICO.
[2] Facultad de Química, Universidad de Guanajuato, Gto, MEXICO.
* To whom correspondence should be addressed: aromo-uribe@fis.unam.mx

ABSTRACT

In this research the thermal and mechanical properties of composites based on recycled high-density polyethylene (HDPE) and recycled Tetrapak have been investigated. The matrix and filler are recovered from landfills. Multicolor HDPE mixtures, with varying concentration of tetrapack flakes, are hot pressed, as well as single color HDPE flakes. Previous studies determine that the nature of the pigment (organics vs. inorganics) strongly influence the mechanical behavior of multicolor HDPE-tetrapack composites. Thus, this research focuses on single color HDPE hot pressed plaques. The kinetics of crystallization under isothermal conditions is determined by differential scanning calorimetry (DSC). The results show that the crystallization kinetics obeys the Avrami theory, and that the Avrami exponent is 1, irrespective of the pigment in use. Small-angle light scattering is applied to investigate the internal structure of the pigmented HDPE. SALS patterns show that the samples exhibited oriented morphologies. However, after melting and slow cooling under pressure the samples exhibit an isotropic morphology. This is confirmed by polarized optical microscopy. Mechanical properties such as Young's modulus, yield stress and ultimate tensile stress are obtained under uniaxial tensile deformation at room temperature. For the single color HDPE plaques the Young's modulus is reduced (after melting), suggesting that the anisotropic molecular chains contribute to the higher value of Young's modulus.

INTRODUCTION

There are worldwide research efforts to reincorporate post-consumer plastics to the production chain, mainly as composite and nanocomposite materials. Due to the versatility, ease of production and recycling ability of PP, new composites are being developed incorporating man-made and natural fibers. For instance, recycled polypropylene (PP) is being reinforced with wood [1], natural spruce fibers [2], with fly ash particles which are a byproduct of thermal power plants [3], and with vegetable fibers [4]. On the other hand, recycled high density polyethylene (HDPE), another commodity thermoplastic is being reinforced with wood [5], finding an optimum concentration of 5% g/g wood before mechanical properties are deteriorated. Moreover, polyvinyl chloride (PVC) has also been recycled and reinforced with acrylonitrile butadiene rubber [6]. The results obtained thus far have shown that these microfillers enhance the rate of crystallization of PP, under isothermal and non-isothermal conditions [7]. Plastic recycling has become a very important activity in Mexico (and worldwide) in the last few years. This is not only due to commercial and economic interest but also to new and stricter environmental regulations in large cities. Polyethylene terephthalate (PET) and high density polyethylene (HDPE) are among the most sought materials for recycling based on the consumption volume that makes economically viable their recycling in Mexico [8]. Most of the post-consumer plastic collected in Mexico is currently separated, pressed, baled and directly shipped overseas.

In our lab we have investigated the production of composites based on recycled HDPE and Tetrapak containers [9,10]. Tetrapak is a composite material widely used in Mexico for milk recipients, and it is not recycled in Mexico because it is a three layer composite based on carton, polymer and aluminum, plus adhesives. In previous studies we have shown that the sort of pigment utilized in HDPE strongly influences the mechanical properties of the composites. Moreover, the degree of crystallinity and mechanical modulus varies among different colored HDPEs suggesting the influence of the pigments to nucleate crystallites. Furthermore, for multicolor HDPE/Tetrapak composites it is found that there is poor adhesion between HDPE samples with different colors, adding to the failure mechanism of the composites.

Therefore, the crystallization kinetics, morphology, and mechanical properties are investigated on recycled single color HDPE molded plaques, with the aim to understand the influence of thermo-mechanical processing on thermal and mechanical properties, thus enabling the production of higher value engineering composites.

EXPERIMENTAL PROCEDURE

Samples. Post consumer HDPE bottles are collected from landfills, and consist of different end use products (i.e., shampoo, motor oil, detergents and so on) thus exhibiting different colors. A typical process of HDPE recycling, starting from selection of material, then crushing it to about ½ inch size flakes, washing in a cold water bath with dilute caustic soda, rinsing with cold water, and dry the product utilizing hot air in columns to eventually collect the material in jumbo bags. The production of molded plaques is carried out placing HDPE flakes in an oven, and holding for 10 min at 200-300°C to allow for HDPE to melt (T_m=125°C). The plaques molding process is conducted by applying 1 Ton pressure to the molten material, thus producing boards approximately 3 mm thick.

Thermal analysis. The thermal behavior is investigated by differential scanning calorimetry (DSC). For DSC experiments the DSC6000™ calorimeter manufactured by Perkin Elmer (Connecticut, USA) is used. Temperature and enthalpy calibration are carried out using analytical grade indium (T_m = 156.6 °C) and corrections are made for the instrument 'baseline'.

Optical microscopy. Polarized optical micrographs are obtained using a Leitz Laborlux D optical microscope. Micrographs are acquired using a Moticam 1000 digital camera (Motic Inc., China), and images are analyzed using ImageTool® v3.0 (The University of Texas Health Science Center -UTHSC- in San Antonio, Texas, USA).

Small-angle light scattering, SALS. Small-angle light-scattering patterns in H_V polarization condition are obtained using an in-house instrument equipped with a He-Ne laser (λ=632.8 nm) and described in detail elsewhere [11]. The light source is a vertically polarized He-Ne laser (wavelength λ=632.8 nm) of 0.8 mW power (model 1500 manufactured by JDS Uniphase Corp., Santa Rosa California, USA). The 0°-90° orientation of the polarizer and analyzer sets the so-called H_V polarization condition. SALS patterns are recorded with a charge-coupled device (CCD, model PC-23C, Super Circuits, Taiwan) and the CCD has a resolution of 200 μm/pixel. The image analysis is carried out with software ImageTool® v.3.

RESULTS AND DISCUSSION

Plaques of single color recycled flakes of HDPE are compression molded in the molten state. Moreover, multicolor HDPE flakes are dry mixed with Tetrapak flakes and compression molded to produce the composite. Figure 1 shows photographs of the molded plaques.

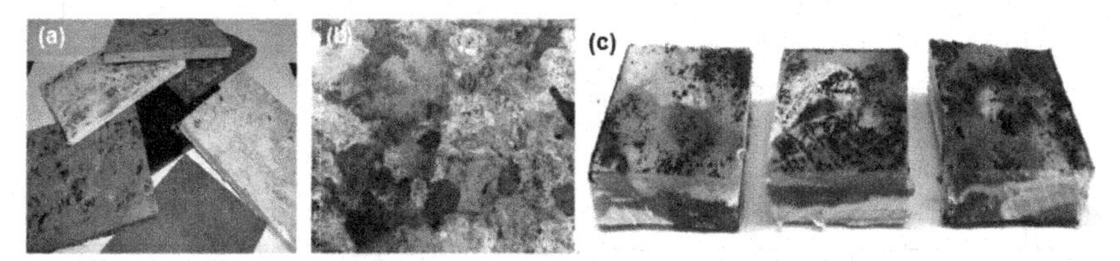

Figure 1. a) Recycled HDPE flakes are compression molded in the molten state. b) Tetrapak flakes are added to the multicolor flakes to produce composites. c) Composites.

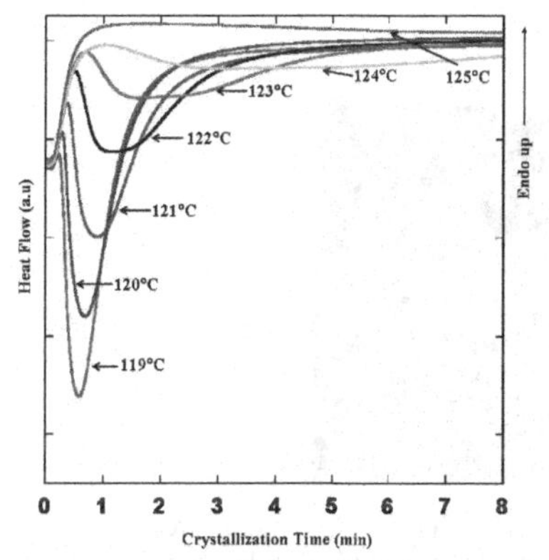

Figure 2. Crystallization exotherms of HDPE, yellow pigmented, carried out isothermally at the indicated temperatures.

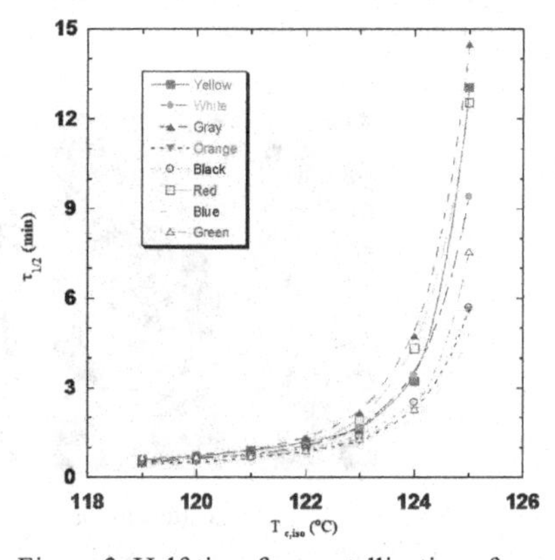

Figure 3. Half-time for crystallization of recycled pigmented HDPEs, as a function of crystallization temperature.

The isothermal crystallization of the single colored HDPEs is investigated by differential scanning calorimetry (DSC). Figure 2 shows typical crystallization exotherms after holding isothermally at the indicated temperatures. The time corresponding to the exotherm minimum is termed the half-time of crystallization, $\tau_{1/2}$. The results of Figure 2 show that $\tau_{1/2}$ increases as the crystallization temperature increases. Figure 3 shows a plot of $\tau_{1/2}$ as a function of crystallization temperature for allthe single colored samples. The results show a distinct difference in rate of crystallization among the different colored HDPEs, the fastest rate occurring for the black pigmented HDPE and the slowest rate for the gray pigmented HDPE, i.e., the nature of the pigment defines the rate of crystallization of HDPE.

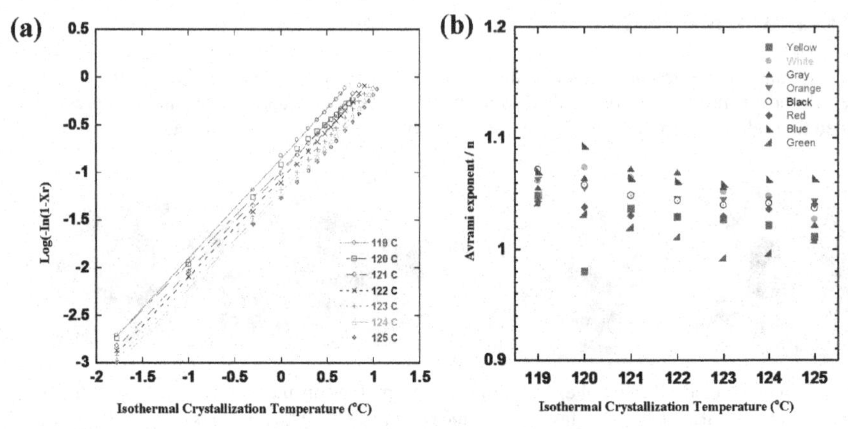

Figure 4 (a) Avrami plot for recycled HDPE, black pigmented. (b) Avrami exponent *n* for recycled single colored HDPEs.

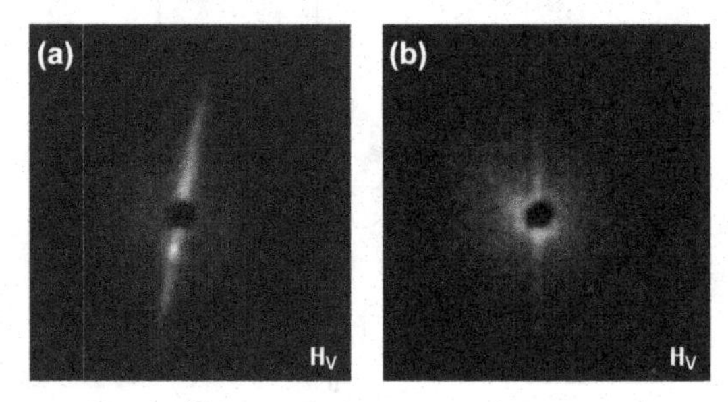

Figure 5. Small-angle light scattering patterns of recycled HDPE, black pigmented. (a) as-molded, and (b) after melted and cooled down to room temperature under pressure. H_V polarization.

The crystallization data is found to conform to the Avrami theory, Figure 4a shows the Avrami plot for the black pigmented HDPE. It can be seen that the experimental data fit very well to a straight line, as predicted by Avrami's theory. From this analysis the Avrami exponent *n* is determined for each colored HDPE, Figure 4b shows the plot of *n* as a function of crystallization temperature and pigment in use. In all cases the exponent *n* is found to be 1.05 ± 0.05, suggesting a linear nucleation mode for crystallization of the HDPEs.

The microstructure of the individual colored HDPEs is investigated by SALS, Figure 5a shows the SALS pattern for the as-molded black pigmented HDPE. The pattern shows a clear anisotropy (diamond-shaped intensity), suggesting alignment of the microstructure due to the applied pressure during molding processing. In the experimental procedure it is described that the molded plaques are quenched rapidly into cold water, and therefore the SALS results suggest that the pressure-induced anisotropy is preserved. Figure 5b shows the SALS pattern corresponding to the same sample after melted and slowly cooled down to room temperature. These results show that the diamond-shaped intensity has been greatly reduced, and there is now amorphous scattering around the beam stop. Therefore, the alignment of the microstructure is greatly reduced for this particular sample. Experiments on other samples show a loss of anisotropy after melting and slow cooling, too.

Table 1. Ultimate tensile stress UTS, Young's modulus E and strain at fracture ε_f of recycled HDPEs: as-molded [a], and melted and cooled down under pressure [b].

HDPE	UTS[a] (MPa)	E[a] (MPa)	ε_f[a] (%)	UTS[b] (MPa)	E[b] (MPa)	ε_f[b] (%)
Yellow	12	700	33	19	430	33
White	8	470	27	15	490	17
Gray	19	1180	17	18	420	47

The mechanical properties under uniaxial tension are determined for the melted and cooled down samples, Table 1 summarizes the results for some of the single colored HDPEs. Data are also included for the as-molded plaques [10]. The results show a significant increase in UTS and ε_f, and sometimes a decrease in E probably due to the loss of anisotropy after melting the as-molded samples.

CONCLUSIONS

Results show that improving processing parameters on recycled HDPE can open up the opportunity to generate high value composite materials with mechanical properties suitable to structural applications. This study also shows that the rate of crystallization is dependent of the pigment utilized. That is, the pigments act as nucleating agents. Moreover, the crystallization kinetics of the recycled HDPEs obeyed the Avrami theory. However, the Avrami exponent is found to be 1, irrespective of the pigment utilized. Small-angle light scattering shows that the as-molded plaques exhibit oriented morphologies. However, after melting and slowly cooling down under pressure the samples exhibit an isotropic morphology. This is confirmed by polarized optical microscopy. The process of melting and cooling down the as-molded single color HDPEs, reduced the Young's modulus (relative to values exhibited by the as-molded plaques), suggesting that the pressure applied for molding induces anisotropy and contributes to the larger Young's modulus.

ACKNOWLEDGMENTS

B Alvarado-Tenorio has been supported by a postdoctoral fellowship from the Mexican Council for Science and Technology, CONACyT. This research has been supported by DGAPA-UNAM, under the PAPIIT program, grant IN109810, and by Nova Polytech SA de CV.

REFERENCES

1. T.S. Khoo, M.M. Ratnam, S.A.B. Shahnaz and H.P.S. Abdul Khalil, J Reinf Plas Compos. 27, 1723 (2008).
2. T. Quynh Truong Hoang, F. Lagattu and J. Brillaud, J Reinf Plas Compos. 29, 209 (2010).
3. K. Das, Ray, K. Adhikary, N.R. Bandyopadhyay, A.K. Mohanty and M. Misra, J Reinf Plas Compos. 29, 510 (2009).
4. A. Ashori and A. Nourbaksh, Waste Management. 30, 680 (2010).
5. Y.H. Cui, J. Tao, B. Noruziaan, M. Cheung and S. Lee, J Reinf Plas Compos. 29, 278 (2010).
6. H Ismail and O.W. Kheong, J Reinf Plas Compos. 27, 1649 (2008).
7. C. Valerio-Cardenas, A. Romo-Uribe, R. Cruz-Silva, L Rejon and R. Saldivar-Guerrero, Emerging Materials Research, 1, 39 (2011).
8. A. Ashori and A. Nourbakhsh, Waste Management. 29, 1291 (2009).
9. A. Romo-Uribe and H.F. Yao, presented at the ACS 242nd National Meeting, Division of Polymeric Materials: Science and Engineering, Denver, Colorado, USA, 2011, (in the press).
10. L. Sanchez-Cadena, B. Alvarado-Tenorio, A. Romo-Uribe, B. Campillo, O. Flores, J. Reinf Plas. Compos. (2013), (in the press).
11. A. Romo-Uribe, B. Alvarado-Tenorio and M.E. Romero-Guzmán, Matter. Rev. LatinAm. Metal. Mat. 30, 190 (2010).

Mater. Res. Soc. Symp. Proc. Vol. 1485 © 2013 Materials Research Society
DOI: 10.1557/opl.2013.253

Effect of boron on the continuous cooling transformation kinetics in a low carbon advanced ultra-high strength steel (A-UHSS)

G. Altamirano[1], I. Mejía[1], A. Hernández-Expósito[2,3], J. M. Cabrera[2,3]
[1] Instituto de Investigaciones Metalúrgicas, Universidad Michoacana de San Nicolás de Hidalgo. Edificio "U", Ciudad Universitaria, Morelia, Michoacán, México.
[2] Departament de Ciència dels Materials i Enginyeria Metal·lúrgica, ETSEIB – Universitat Politècnica de Catalunya. Av. Diagonal 647, Barcelona, Spain.
[3] Fundació CTM Centre Tecnològic, Av. de las Bases de Manresa, 1, Manresa, Spain.

ABSTRACT

The aim of the present research work is to investigate the influence of B addition on the phase transformation kinetics under continuous cooling conditions. In order to perform this study, the behavior of two low carbon advanced ultra-high strength steels (A-UHSS) is analyzed during dilatometry tests over the cooling rate range of 0.1-200°C/s. The start and finish points of the austenite transformation are identified from the dilatation curves and then the continuous cooling transformation (CCT) diagrams are constructed. These diagrams are verified by microstructural characterization and Vickers micro-hardness. In general, results revealed that for slower cooling rates (0.1-0.5 °C/s) the present phases are mainly ferritic-pearlitic (F+P) structures. By contrast, a mixture of bainitic-martensitic structures predominates at higher cooling rates (50-200°C/s). On the other hand, CCT diagrams show that B addition delays the decomposition kinetics of austenite to ferrite, thereby promoting the formation of bainitic-martensitic structures. In the case of B microalloyed steel, the CCT curve is displaced to the right, increasing the hardenability. These results are associated with the ability of B atoms to segregate towards austenitic grain boundaries, which reduce the preferential sites for nucleation and development of F+P structures.

INTRODUCTION

The advanced ultra-high strength steels (A-UHSS) such as dual phase (DP), complex phase (CP), boron steels (BS) and martensitic steels (MART) are taking a strong interest in various industrial sectors, particularly the automotive industry. These latest generation steels with multiphase microstructures are characterized by an excellent combination of high strength, good toughness and ductility [1]. However, in order to ensure the industrial application is necessary to know important aspects such as its phase transformation kinetics under non-equilibrium conditions. The overall CCT kinetics can be readily described by the CCT phase diagrams [2]. This kind of diagrams provides precise information on the nature of microconstituents formed from the anisothermal decomposition of austenite. The formation of each transformation product is mainly described by the transformation-start temperature and by the formation cooling rate range. In the practical situation these diagrams are mainly used to predict the microstructure and hence the mechanical properties of steel after thermal and/or thermo-mechanical treatments [3-4]. The chemical composition of steel is considered the most important factor governing the microstructures resulting during the non-equilibrium transformation of austenite [2]. Practically, all alloying elements in the steel tend to retard the austenite decomposition [5]. In the case of the transformations caused by nucleation and growth (ferrite and pearlite), this delay is due to the need to diffuse not only carbon but all elements in solid solution in austenite. Alloying elements such as Mo, Nb, V and Ti tend to retard the formation of ferrite by a solute drag effect, increasing in this way the steel hardenability [6]. Similarly, it is known that B atoms segregate towards austenite grain boundaries (GB) and increase the hardenability by suppressing the nucleation of ferrite [7]. There are four mainly explanations about the B effect on the steel hardenability [7]: (i) B segregation to austenite GB reduces the grain boundary energy, and so the number of preferential nucleation sites for ferrite, (ii) B reduces the self-diffusional coefficient of iron at GB and decreases the nucleation rate of ferrite, (iii) GB are a preferential nucleation sites for ferrite and when B segregates to GB, these sites will vanish, and (iv) Fine borides form along the GB and are coherent with the matrix; in this case, is hard to nucleate ferrite at the boride-matrix interface. Nowadays, in the context of low carbon advanced high strength steels, there is not relevant information strictly focused on the CCT kinetics. Therefore, the aim of this research work is to study the effect of B addition on the CCT kinetics of a new family of low carbon advanced ultra- high strength (A-UHSS) NiCrVCu steels.

EXPERIMENTAL DETAILS

The experimental A-UHSS NiCrVCu steels are melted in the Foundry Laboratory of the Metallurgical Research Institute-UMSNH (México) using high purity raw materials in a 25 kg capacity induction furnace and cast into 70 mm × 70 mm cross section ingots. Table I shows the chemical composition of the two experimental steels examined in this study.

Table I. Chemical composition of the studied A-UHSS steels (wt. %).

A-UHSS steel	C	Mn	Si	S	Cu	Cr	Ni	V	Al	N	B
B0	0.15	0.40	0.42	0.02	0.52	1.31	2.44	0.22	0.0026	0.0091	**0**
B5	0.09	0.41	0.29	0.02	0.50	1.30	2.42	0.22	0.0031	0.0087	**0.0214**

Tensile tests revealed that in the as hot-rolled + quenched condition the ultimate tensile strength (UTS) of B0 and B5 steels are 925 and 1135 MPa, respectively, with complex phases consisting mainly of bainite and martensite [8]. The dilatometric study is carried out in a Bähr DIL 805 quenching and deformation dilatometer. Cylindrical specimens of 4 mm in diameter and 10 mm in length are machined from the as hot-rolled + quenched condition steels. First, to determine the critical temperatures (Ac_1, Ac_3 and M_s) samples are austenitized at 1050 °C for 5 min followed of quenched with helium gas. These temperatures indicate the start and finish of the austenite transformation and the start of the martensitic transformation, respectively. The thermal cycle used for the construction of the CCT diagrams is: (i) heating at 5 °C/s from room temperature up to austenitization temperature of 1050 °C for 5 min, (ii) cooling at 2 °C/s up to 900 °C for 1 min, and (iii) continuous cooling at different rates (0.1, 0.25, 0.5, 1, 2.5, 5, 10, 20, 50, 100 and 200 °C/s). Finally, the constructed CCT diagrams are verified by metallographic characterization using optical microscopy (LOM) and Vickers micro-hardness tests.

DISCUSSION

Dilatometric curves

Figures 1and 2 shows all the dilatation curves obtained for the studied steels. As can be noticed, the anisothermal decomposition of austenite depends on the cooling rate.

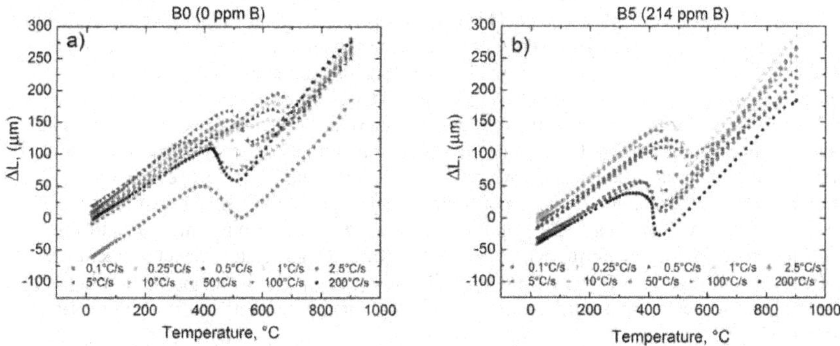

Figure 1. Dilatometric curves obtained at different cooling rates. a) B0 steel (0 ppm B), and b) B5 steel (214 ppm B).

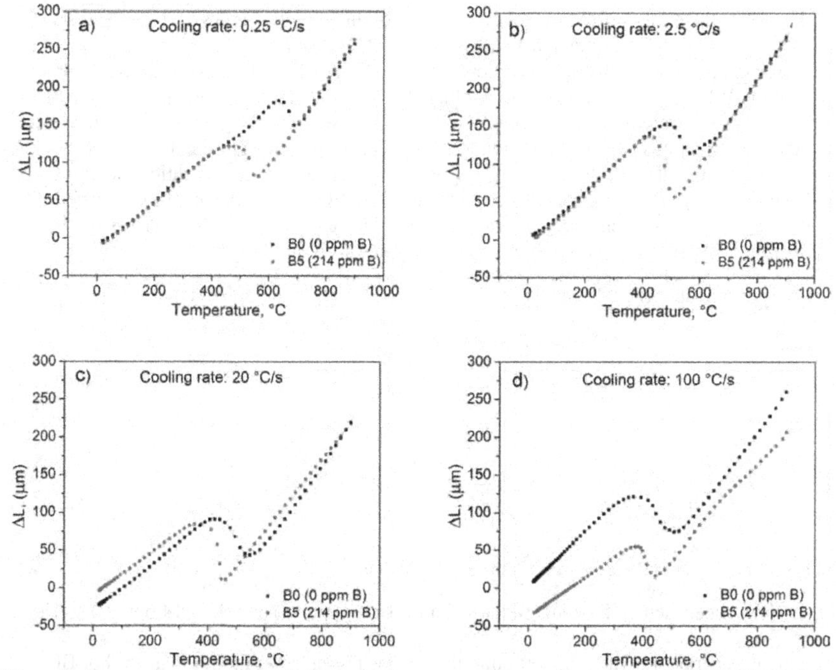

Figure 2. Comparison of obtained dilatometric curves of the studied steels. a) 0.25 °C/s, b) 2.5 °C/s, c) 20 °C/s and d) 100 °C/s.

The results of tests conducted to determine the critical transformation temperatures showed lower values in the non-microalloyed steel. Temperatures of 725, 826, and 413 °C are recorded for Ac_1, Ac_3 and M_s, respectively, for the case of B0 steel; while temperatures of 747, 852 and 463 °C are recorded for B5 steel. This behavior is related to the higher carbon content in B0 steel (see table I). As is well-known [5], carbon is a strong austenite stabilizer, therefore, the higher the carbon content the lower the critical transformation temperatures Ac_1 and Ac_3. Similarly, increasing the carbon concentration decreases the martensite-start temperature (M_s). In general, it is observed that the transformation temperature of austenite decreases as cooling rate increases. As is well-known [2], the sequence of phase transformations in steels by continuous cooling from the austenite is ferrite, pearlite, bainite and martensite. Therefore, each slope variation detected in these dilatometric curves corresponds to a specific phase transformation. On the other hand, it is evident from figure 2 that B additions delay the decomposition kinetics of austenite to ferrite. Phase transformations starts and finishes at significantly lower temperatures in the B5 steel. This trend is observed for nearly all cooling rates. This behavior has already been reported in bainitic steels and is attributed to B segregation and precipitation of fine borocarbide at austenite GB [9].

CCT diagrams, microstructures and hardness results

The CCT diagrams constructed by the analyses of dilatometric curves with the Vickers micro-hardness results are illustrated in figure 3. The CCT diagram for B0 steel (figure 3a) exhibits three mains regions corresponding to the ferrite (F), pearlite (P) and bainite (B) phase transformation. When the cooling rate is lower than 2.5 °C/s coexists ferrite and pearlite phases resulting in the lowest micro-hardness values (214-246 HV), whereas the presence of bainite and a small amount of martensitic transformation led to an increased hardness of up to 339 HV. According to the CCT diagram for B5 steel (figure 3b), it is clear that the B addition increases the steel hardenability. The CCT curves are shifted to the right, opening the metastable austenite field thus increasing the hardenability. As can be seen in this CCT diagram, the martensitic transformation take place at slower cooling rates compared with the B0 steel.

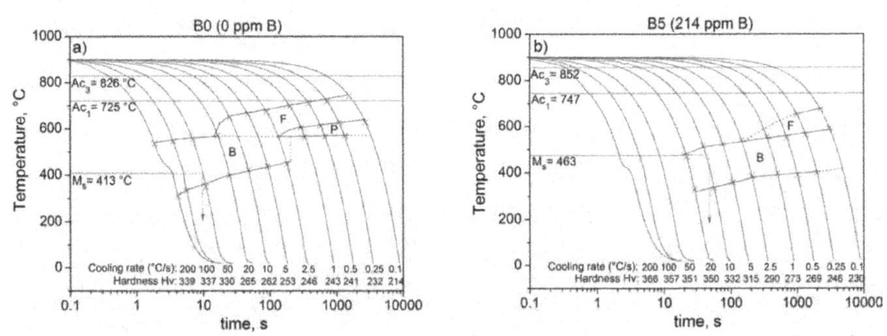

Figure 3. CCT experimental diagrams constructed for B0 (0 ppm B) and B5 (214 ppm B) steels.

The microstructural validation of both diagrams is presented in figures 4 and 5. For B0 steel at the slowest cooling rates, the microstructure is a mixture of ferrite and pearlite (F+P) phases in different proportions, depending on the cooling rate (see figures 4a-c). At 10 °C/s (figure 4d) the microstructure is predominantly bainitic, while a microstructure consisting of bainite and martensite (B+M) is observed from cooling rate of 50 °C/s (see figures 4e-f).

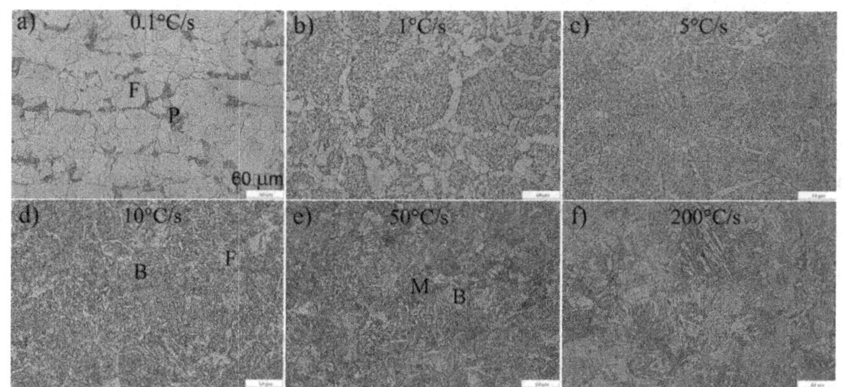

Figure 4. Optical microstructures for B0 steel obtained after different cooling rates. a) 0.1°C/s, b) 1°C/s, c) 5 °C/s, d) 10 °C/s, e) 50 °C/s and f) 200 °C/s.

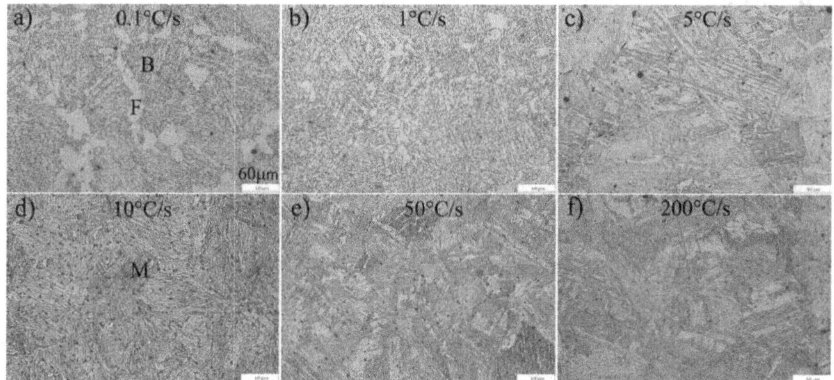

Figure 5. Optical microstructures for B5 steel obtained after different cooling rates. a) 0.1°C/s, b) 1°C/s, c) 5 °C/s, d) 10 °C/s, e) 50 °C/s and f) 200 °C/s.

As shown in figure 5, for the same cooling rates (0.1-5 °C/s) B5 steel exhibits a microstructure mainly composed by bainite and some ferrite that tends to disappear as the cooling rate is increased. Unlike B0 steel, the martensitic transformation in B5 steel is observed from 10 °C/s, which is consistent with the CCT diagrams. This behavior is explained mainly in terms of B segregation to austenite GB, which retards or inhibits the nucleation of ferrite, increasing in this way the steel hardenability [7]. Finally, for all cooling rates the hardness test results revealed higher values in the B microalloyed steel (B5) (see figure 3). These results can also be related with a precipitation hardening effect. Supersaturated B can precipitate as iron-boride, Fe_2B and/or iron-borocarbide, $Fe_{23}(B, C)_6$, during the cooling, as it is shown by Linier et al. [10].

CONCLUSIONS

It is found from continuous cooling treatments that the B addition in a low carbon advanced ultra- high strength steel (A-UHSS) delays the decomposition kinetics of transformation of austenite to ferrite, thereby promoting the formation of bainitic-martensitic structures. The B addition to A-UHSS shifts the CCT curves to the right increasing the hardenability of steel. This effect is associated to segregation of B atoms towards austenitic GB, which reduce the preferential sites for nucleation and development of ferritic-pearlitic (F+P) structures.

ACKNOWLEDGMENTS

G. Altamirano would like to thank CONACYT (México) for the scholarship support during this project. Authors also acknowledge CMEM-UPC (Spain), for the support and technical assistance in this research work. Funding is obtained through project CICYT-MAT2008-06793-C02-01 (Spain) and CIC-UMSNH (México).

REFERENCES

1. Committee on Automotive Applications, International Iron & Steel Institute, *Advanced High Strength Steel Application Guidelines*, pp. 1–13 (2009).
2. J. C. Zhao and M. R. Notis, *Mater. Sci. Eng. Rep.* **15**, 135-207 (1995).
3. A. B. Cota, P. J. Modenesi, R. Barbosa and D. B. Santos, *Scripta Mater.* **40**, 165-169 (1999).
4. W. You, W. H. Xu, Y. X. Liu, B. Z. Bai and H. S. Fang, *J. Iron Steel Res. Int.* **14**, 39-42 (2007).
5. H. K. D. H. Bhadeshia and R. W. K. Honeycombe, *Steels, Microstructures and Properties,* third edition (Elsevier, Butterworth-Heinemann, 2006), pp. 71-92.
6. C. Zhang, D. Cai, Y. Wang, M. Liu, B. Liao and Y. Fan, *Mater. Charact.* **59**, 1638-1642 (2008).
7. X. M. Wang and X. L. He, *ISIJ Int.* **42**, 38-46 (2002).
8. I. Mejía, A. García de la Rosa, A. Bedolla-Jacuinde and J. M. Cabrera, *Mater. Res. Soc. Symp. Proc.* **1373**, 89-94 (2012).
9. K. Zhu, C. Oberbilling, C. Misik, D. Loison and T. Iung, *Mater. Sci. Eng.* **A528**, 4222-4231 (2011).
10. L. Lanier, G. Metauer and M. Moukassi, *Mikrochim. Acta* **114-115**, 353-361 (1994).

Mater. Res. Soc. Symp. Proc. Vol. 1485 © 2013 Materials Research Society
DOI: 10.1557/opl.2013.254

Effect of mechanical activation on the crystallization and properties of iron-rich glass materials

Claudia M. García-Hernández, Jorge López-Cuevas and José L. Rodríguez-Galicia
Centro de Investigación y de Estudios Avanzados del IPN, Unidad Saltillo, Carr. Saltillo-Monterrey, Km. 13.5, C.P. 25900, Ramos Arizpe, Coahuila, México.

ABSTRACT

The effect of mechanical activation (MA) of the precursor mixture of raw materials and/or the parent glass, on the microstructure and physical and mechanical properties of iron-rich glass-ceramic materials of the system SiO_2-B_2O_3-BaO-Fe_2O_3, has been studied. MA of the materials is conducted for 0, 2 or 6h using a high energy attrition milling device. Crystallization treatments are given to the parent glass at 650, 750 or 850°C for 5h. Crystallization of the samples is promoted by increased treatment temperature, and especially also by double MA at 850°C. With increasing crystallization temperature, both the density and the compressive strength increase, while porosity decreases. However, at 850°C, prolonged MA decreases both the density and the compressive strength due to an increment in porosity caused by the growth of the $BaFe_{12}O_{19}$ crystals.

INTRODUCTION

Mechanical activation (MA) is a process that involves prolonged high energy milling of a material. Its use can lead to many interesting applications, ranging from waste processing to the production of advanced materials. During MA, several structural changes take place in the materials being ground, which modify their physicochemical properties [1]. An important feature of this processing route is the refinement of the microstructure (grain and particle size) associated with the particle fracture and deformation caused by the collision events occurring between the grinding medium and the powder. It can also increase the reactivity of the milled powders, leading to a reduction in the temperature conventionally required to carry out a posterior solid state reaction process for the synthesis of a material with particular chemical and phase composition. MA has been widely associated with the amorphization of crystalline materials, but its use to promote the crystallization of amorphous materials is a less studied case [2-3].

The main aim of this work is to study the effect of MA of the precursor mixture of raw materials and/or the parent glass, on the microstructure, and physical and mechanical properties of iron-rich glass-ceramics of the system B_2O_3-SiO_2-BaO-Fe_2O_3, for potential structural applications, intending to obtain in them barium hexaferrite ($BeFe_{12}O_{19}$) crystals dispersed in a borosilicate glass matrix. High purity raw materials as well as an iron oxide husk are used to obtain the desired glass and glass-ceramic materials. The iron oxide husk is a byproduct of the knurling of the surface of low carbon steel tubes prior to their coating with a refractory lining during the production of oxygen lances for steel making.

EXPERIMENTAL DETAILS

The studied parent glass is based on one of the materials investigated by A. Mirkazemi et al. [4]. Its chemical composition is (wt.%): 4.63% SiO_2, 10.74% B_2O_3, 41.45% BaO and 43.16% Fe_2O_3. The stoichiometric mixture of raw materials is composed by (wt.%): 3.0% SiO_2, 16.0% H_3BO_3, 44.68% $BaCO_3$ and 36.35% of iron oxide husk (90.77% Fe_3O_4, 6.63% FeO(OH) and 2.48% SiO_2). The latter material is supplied by SIRSSA, Guadalupe, N.L., México. It is analyzed by semiquantitative short wavelength dispersive X-Ray Fluorescence (XRF) spectroscopy, using a S4 PIONNER BRUKER apparatus. Prior to their analysis, the samples are milled to a particle size of ~105 μm using a planetary mill with agate balls and containers, which is followed by uniaxial pressing employing wax as lubricant. A 71-element scanning (from Na to U) is carried out, with a total of 588 lines, under the following conditions: Vacuum-sealed XR tube fitted with Rh anode, accelerating voltage of 25 to 60 kv, 0.46dg collimator, 34mm collimator mask, and LIF200 (lithium fluoride) and PET (pentaerithrit) crystals.

To obtain the parent glass, the initial mixture of raw materials is melted in mullite ($Al_6Si_2O_{13}$) crucibles, using a gas welding torch at ~2000°C, which is followed by fast cooling in ice water in order to prevent the crystallization of the glass. Once obtained, in order to produce a homogeneous initial particle size of ~105 μm, the parent glass is milled for 5 min. in a planetary mill using agate balls and containers. The MA of the mixture of raw materials and/or the parent glass is conducted for 0, 2 or 6h in a Teflon-lined closed chamber laboratory attrition mill operated at 1700 rpm, employing 8 mol.% MgO-partially-stabilized ZrO_2 balls with a diameter of 1mm as milling media, and with a mass ratio of 5:1 with respect to the mill load; ethylic alcohol is added as dispersion medium until 3/4 of the volume of the mill container is filled. Hereinafter, the MA treatment given to the materials is denominated as: a) 0-0 for the glass-ceramic materials synthesized without any kind of MA; b) 2-0 or 6-0 for samples with either 2h or 6h of MA given to the mixture of raw materials and without MA of the parent glass; c) 0-2 or 0-6 for samples without MA of the mixture of raw materials and with either 2h or 6h of MA given to the parent glass, and d) 2-2 or 6-6 for samples with either 2h or 6h of MA given to both the mixture of raw materials and the parent glass. Prior to the crystallization treatment, the parent glass, with or without MA, is uniaxially pressed applying a load of 5-6 Ton in order to obtain 2g-cylinders with a height of 5 mm and a diameter of 12 mm. The glass-ceramic materials are then obtained by heat treatment of the parent glass at temperatures of 650, 750 or 850°C for 5h, using a Thermolyne F62700 electric furnace. The glass and glass-ceramic materials are analyzed by X-Ray Diffraction (XRD) using a Philips X-PERT equipment and Cukα radiation. The microstructure of graphite-coated glass and glass-ceramic materials is observed using a Philips XL30 ESEM Scanning Electron Microscope (SEM), employing an acceleration voltage of 20-30 kV and the Backscattered Electron Imaging (BEI) mode, with a working distance of 10 mm. Prior to this, the specimens are ground and polished using standard ceramographic techniques, and then chemically attacked for 10s with 10 vol.% Nital reactant. In order to determine the real density (ρ_{Real}, g/cm^3) of the glass-ceramic materials, in such a way as to avoid the effect of porosity, they are first milled by using diamonite mortar and pestle, until a particle size of ~60 μm is obtained. Then, the density of the powdered materials is determined by the Archimedes' principle using 1g-samples and a 25 ml-glass pycnometer, employing toluene as liquid medium. The apparent density (ρ_{App}, g/cm^3) of the monolithic glass-ceramics is determined by the Archimedes' principle in distilled water at room temperature. The relative density (ρ_R, %) of the glass-ceramics is determined from the ρ_{App}/ρ_{Real} ratio. Total porosity of the glass-ceramic materials is determined from the measured densities, according to the ASTM C20-00 standard [5]. Their compressive strength is determined using an automated hydraulic 50-C702 CONTROLS machine with a cell capacity of 15 KN; the application load rate is 10 MPa/s.

DISCUSSION

Characterization of the parent glass

The XRD and SEM analyses verify that the parent glass is essentially amorphous, with the presence of a very small amount of tiny Fe_3O_4 crystals. Microarea SEM/EDS analysis shows that the experimental chemical composition of the parent glass is very close to the nominal one.

Effect of mechanical activation on the phase evolution of the synthesized glass-ceramic materials

The XRD patterns of Figures 1 and 2 show the presence of an amorphous hump centered at an angle 2θ of ~30°, denoting an incomplete crystallization of the heat-treated samples.

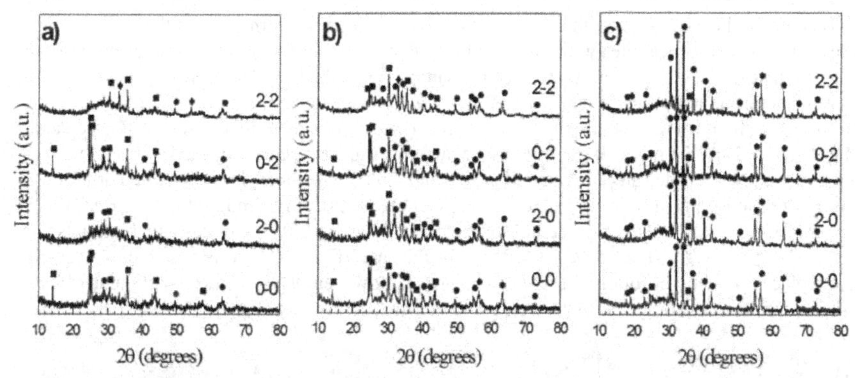

Figure 1. XRD patterns of glass-ceramics synthesized with or without MA given to the precursor mixture of raw materials and/or to the parent glass for 2h, and heat treatment carried out at: a) 650°C, b) 750°C, and c) 850°C. Key: •$BaFe_{12}O_{19}$, ▪$Ba(B_2O_4)$, and ♦Fe_3O_4.

For all sintering temperatures, the MA given to the mixture of raw materials, either for 2h or 6h, is found to be detrimental to the crystallization of the glass-ceramic materials, without any significant differences existing between both milling times. However, this effect becomes less important with increasing sintering temperature. The BaB_2O_4 phase predominates in the samples sintered at 650°C, while $BaFe_{12}O_{19}$ is by far the most abundant crystalline phase formed at 850°C. Both phases have comparable relative proportions at 750°C, with the BaB_2O_4 phase showing a slightly higher concentration at this temperature.

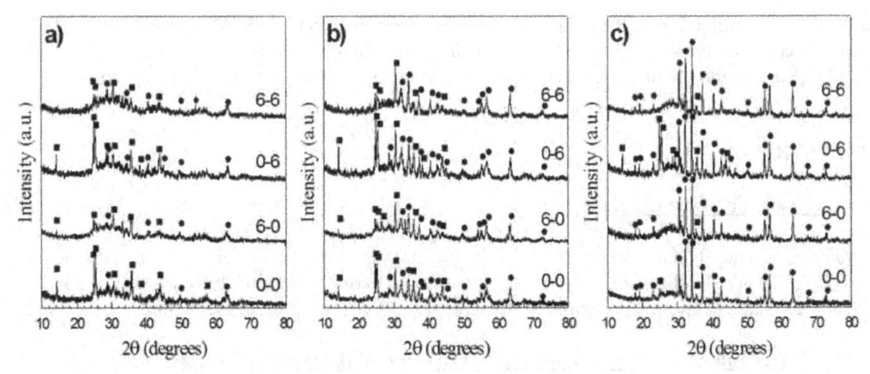

Figure 2. XRD patterns of glass-ceramics synthesized with or without MA given to the precursor mixture of raw materials and/or to the parent glass for 6h, and heat treatment carried out at: a) 650°C, b) 750°C, and c) 850°C. Key: •$BaFe_{12}O_{19}$, ▪$Ba(B_2O_4)$, and ♦Fe_3O_4.

In contrast, the MA given to the parent glass slightly favors the formation of the $BaFe_{12}O_{19}$ phase in the case of the samples with the 0-2 MA, and strongly favors the formation of both the $BaFe_{12}O_{19}$ and BaB_2O_4 phases in the case of the samples with the 0-6 MA, especially with increasing sintering temperature. For the 0-2 MA, for all sintering temperatures, the relative proportions of the formed crystalline phases behave in a similar way to that observed for the case of the samples with MA given to the mixture of raw materials. In the case of the 0-6 MA, the same thing applies for the samples sintered at 650 and 750°C. However, for this particular MA the relative proportions of the $BaFe_{12}O_{19}$ and BaB_2O_4 phases are very similar at 850°C. Lastly, the double MA treatments are detrimental for the crystallization of the glass-ceramic materials that are heat-treated at 650 or 750°C, especially for the case of the 2-2 MA, in which the Fe_3O_4 phase replaces partially or totally the $BaFe_{12}O_{19}$ phase. However, at 850°C these particular MA conditions produce the largest observed formation of the $BaFe_{12}O_{19}$ phase.

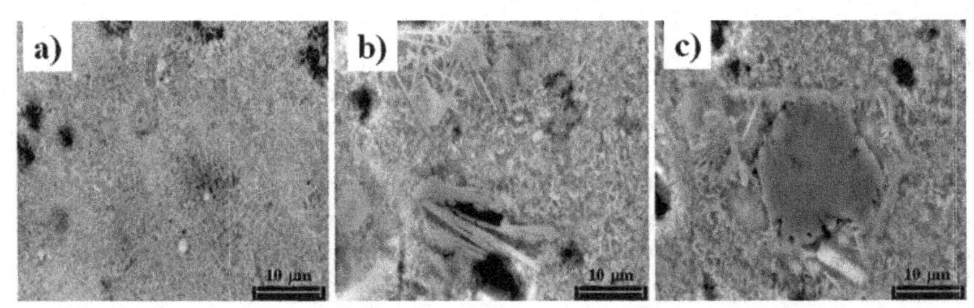

Figure 3. BEI-SEM micrographs of glass-ceramic materials synthesized at 850°C. MA treatment: a) 0-0, b) 2-2, and c) 6-6.

The SEM micrographs of Figure 3, corresponding to glass-ceramics sintered at 850°C, show that the $BaFe_{12}O_{19}$ crystals change from small needles with a size of ~ 1-2 μm for the material with the 0-0 MA, to needles with similar size plus platelets with a size of ~ 10-20 μm and blocky crystals with a size of ~ 5-25 μm for the materials with the 2-2 MA and the 6-6 MA treatments.

Effect of mechanical activation on the physical and mechanical properties of the synthesized glass-ceramic materials

Density and Porosity

Figures 4a and 4b show that, for all sintering temperatures, the glass-ceramic materials having the highest relative densities and the lowest total porosities are obtained with MA given to the mixture of raw materials for 2h. It is noticeable that this corresponds to the least crystallized samples. Thus, it seems that the advance of crystallization increases porosity in the glass-ceramic materials, which is likely associated with the growth of the $BaFe_{12}O_{19}$ crystals. Probably the lowest relative density and the highest total porosity are obtained for the glass-ceramic materials with MA given to the raw material mixtures for 6h due to a more pronounced growth of the $BaFe_{12}O_{19}$ crystals taking place under such conditions.

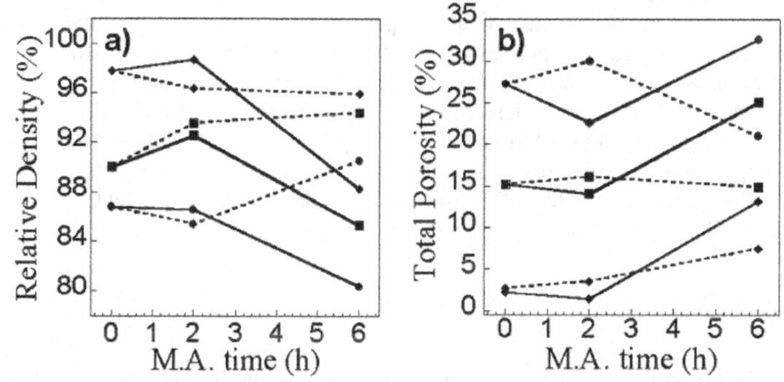

Figure 4. a) Density and b) Porosity of glass-ceramic materials synthesized at •650°C, ■750°C or ♦850°C, with different MA treatments. (—) Mixture of raw materials, and (····) Parent glass.

Compressive Strength

Figures 5a and 5b show that the highest values of compressive strength are obtained for the glass-ceramic materials synthesized at the highest sintering temperature employed (850°C). However, at the latter temperature, for the case of the materials obtained with MA given for 6h either to the mixture of raw materials or to the parent glass, the compressive strength drops significantly due to the high porosity present in the glass-ceramics, which is probably caused by a faster growth of the $BaFe_{12}O_{19}$ crystals under those conditions. The maximum value obtained for the compressive strength is 387.20 ± 4.16 MPa, which corresponds to a glass-ceramic material synthesized with the 0-2 MA and heat treatment at 850°C.

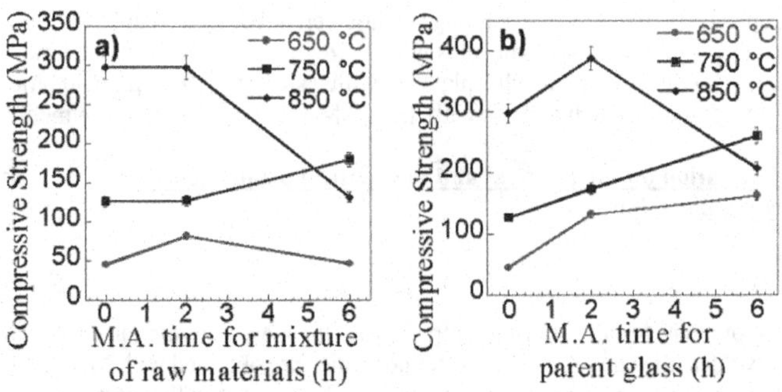

Figure 5. Compressive strength vs MA time for: a) Mixture of raw materials and b) Parent glass.

CONCLUSIONS

Glass crystallization is promoted by increasing temperature, with or without MA. However, full crystallization is never attained under the investigated experimental conditions. MA of the precursor mixture is detrimental to crystallization, especially at low sintering temperatures. In contrast, MA of the parent glass favors crystallization, especially with increasing sintering temperature for the 0-6 MA. The double MA is detrimental for crystallization at 650 or 750°C, but at 850°C the 2-2 MA produced the largest formation of $BaFe_{12}O_{19}$ crystals. Both the density and the compressive strength increase and porosity decreases with increasing crystallization temperature. The first two properties decrease with 6h of MA due to the porosity increment caused by the growth of the $BaFe_{12}O_{19}$ crystals.

ACKNOWLEDGMENTS

The authors express their gratitude to CONACYT and Cinvestav-Saltillo for the financial support and facilities provided for the development of this work.

REFERENCES

1. M. Vlahovic, S. Martinovic, P. Jovanic, T. Boljanac and V. Vidojkovic in *Mechanochemical activation of mixtures for low-melting glasses production*, (Proc. Eur. Cong. Chem. Eng. ECCE-6, Copenhagen, 2007) pp. 16-20.
2. P.G. McCormick, and F.H. Froes in *The fundamentals of mechanochemical processing*, (The Brimacombe Mem. Symp., 2000)
3. C. Suryanarayana, *Mechanical alloying and milling*, (Marcel Dekker, 2004).
4. M. Mirkazemi, V.K. Marghussian and A. Beitollahi, *Ceram. Int.* **32**, 43-51 (2006).
5. A.S.T.M C20-00, Standard test methods for apparent porosity, water absorption, apparent specific gravity, and bulk density of burned refractory brick and shapes by boiling water.

Mater. Res. Soc. Symp. Proc. Vol. 1485 © 2013 Materials Research Society
DOI: 10.1557/opl.2013.255

EFFECTS OF HOLDING TIME ON HAZ-SOFTENING IN RESISTANCE SPOT WELDED DP980 STEELS

C.J. Martínez-González[1,2], A. López-Ibarra[1], S. Haro-Rodriguez[1], V.H. Baltazar-Hernandez[1], S.S. Nayak[2], Y. Zhou[2]

[1]Unidad Académica de Ingeniería, Universidad Autónoma de Zacatecas, México
[2]Mechanical and Mechatronics Engineering, University of Waterloo, Canada

ABSTRACT

Resistance spot welding (RSW) of dual-phase (DP) steel subjected to various conditions of cooling rate (holding time) is studied in this work. Lap-shear tensile testing is used in order to evaluate the mechanical performance of the weldments. The microstructure is analyzed through optical and electron microscopy and the hardness is obtained through Vickers method. Results indicate an effective tempered region along the sub-critical heat affected zone in all the samples. A broken morphology accompanied with presence of small carbides within tempered martensite phase is consistently observed. Variations in the cooling rate (holding time) indicate minimal effect on the degree of softening and on the mechanical performance of the welds.

INTRODUCTION

The demand for advanced high strength steel (AHSS) sheets having a combination of strength and ductility has increased in the automotive industry. Dual-phase (DP) steel is part of the family of AHSS and is composed of a ferritic matrix with varied volume fractions of martensite phase. In particular, DP has become popular in auto-body constructions due to their excellent properties and microstructure [1].

Resistance spot welding (RSW) is the dominant process in the automobile maker facilities. RSW is well known being a fast welding process i.e. it involves rapid heating followed by fast cooling rates. An important parameter in the RSW practice is the holding time. The holding time parameter is associated to the time in which both electrodes remain in contact with the specimen just after welding current has been switched off; in fact, the holding time parameter is directly related to the cooling rate of the specimen [2, 3]. Furthermore, RSW is widely utilized on joining DP steels. An interesting issue in RSW of DP steel is a phenomenon known as heat affected zone (HAZ) softening that involves reduction of hardness at the subcritical-HAZ (SCHAZ) with respect to base metal (BM) [4, 5]. HAZ-softening actually promotes localized strain at SCHAZ along with earlier failures [6]. The reason for the HAZ-softening in RSW of DP steel welds has been attributed in fact to tempering of the martensite phase [7, 8]. Tempering of martensite, which occurs at temperature at or below the lower critical (Ac_1) line, has been reported to be the reason for HAZ-softening in DP steels [9-12]. The degree or severity of tempering is highly dependent on the maximum (peak) temperature and the time at peak temperature during isothermal heat treatments [13, 14].

Even though martensite tempering has been extensively investigated on martensitic steels subjected to isothermal heat treatments, little work has been addressed on investigating the tempering of martensite particles on DP steel during fast thermal cycles (far from equilibrium) like those generated by the RSW process [15]. Furthermore, there is no available literature regarding to the effects of cooling rates (i.e. like those generated by RSW through holding time parameter) on the tempering of martensite (softening) in DP steel.

In this work, RSW of DP steel subjected to various conditions of holding time (i.e. a range starting from 0 to 80 cycles) is studied. The main purpose is to evaluate the effects of cooling rate on the tempering of martensite (softening) on the DP microstructure.

EXPERIMENTAL PROCEDURE

In this work, a DP steel having 980 MPa of UTS is studied. DP steel comprises a chemical composition of 0.13 C, 1.91 Mn, 0.03 Si, 0.01 P, 0.005 S, 0.33 Mo and 0.16 Cr (wt.%), and the carbon equivalent based on Yurioka's equation is of (CE) = *0.391* [16]. DP steel had a thickness of 1.2 mm with hot-dip galvanized coating. Figure 1 shows the microstructure of DP980 steel composed by ferrite (α-matrix) along with banded islands of martensite (α'). The estimation of 48% volume fraction of martensite phase is obtained by standard metallographic techniques.

All welds is conducted using RSW machine, pedestal type, 250 kVA single phase AC, pneumatically controlled, operated with a current of 60Hz. According to the resistance welding manufacturing alliance [17] (RWMA) electrodes used in RSW is female, type dome FB25 with a 6-mm face diameter.

A constant flow water of 4 l/min is maintained for cooling the electrodes according to AWS standards [18]. A stabilization procedure is carried out for the new electrodes to break in contact tips suggested by AWS. The squeeze time and weld time is maintained constant of 25 and 20 cycles respectively, the force of 4.5 kN, the current of 9 kA. The hold time is varied from 0 to 80 cycles. Figure 2 shows the welding schedule used in this work.

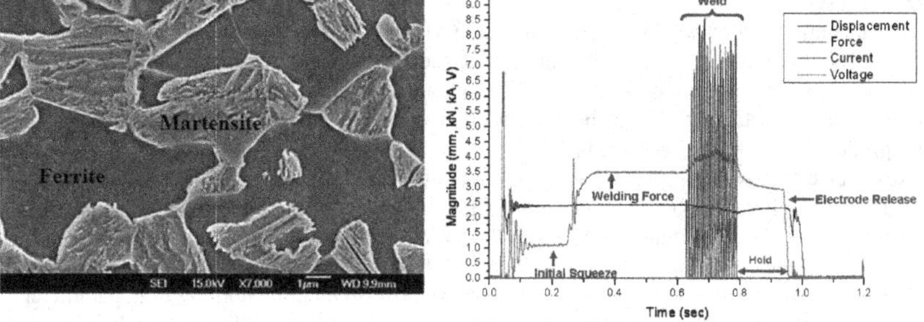

Figure 1. Base metal microstructure of DP980 by SEM.

Figure 2. Welding schedule in RSW

Lap shear tension test is conducted to obtain information about the mechanical performance of the welds. Coupon dimensions for lap shear are 105 x 45 mm. The samples are prepared following AWS standards [18]. The tensile test is carried out with an Instron 4206 universal testing machine. All the tests are performed at room temperature. Testing is performed by employing seven samples for each condition of holding time; the crosshead velocity is constant rate of 10 mm/min^{-1}. Two shims of same thickness are used to maintain the alignment in all samples under shear tensile testing. A schematic of test is shown in figure 3.

Samples are sectioned across the weldment covering various zones (i.e. fusion zone, heat affected zone (HAZ) and base metal (BM)). Sectioned samples are mounted in epoxy resin followed by mechanically polishing using a series of SiC papers; just after previous procedure, fine polishing is performed by employing diamond suspensions (1 μm and 0.25 μm). Further, etching is done with Nital solution of 2% in order to reveal the microstructure (i.e. weld nugget boundary and fusion line). The microstructure is examined by optical microscopy and scanning electron microscopy (SEM). Vickers micro-hardness measurements are obtained of cross-section samples using 200 g load with 15 s of time. Five indentations for each side of the weld is spaced 0.2 mm apart to avoid their interference, the hardness profile is a line 100 μm to Ac_1 (line of critical temperature). The hardness profile is obtained at room temperature using a Vickers microhardness tester according to AWS standards [18].

Peel testing is used to measure the diameter weld size according to AWS standards [18]. Figure 4 illustrates a schematic of the peel testing method. Additionally, measurements are obtained through metallographic techniques. A stereo-microscope is used to measure the diameter of weld size in the cross-sections specimens. The weld size in both techniques is reported as an average of the minimum and maximum nugget dimensions.

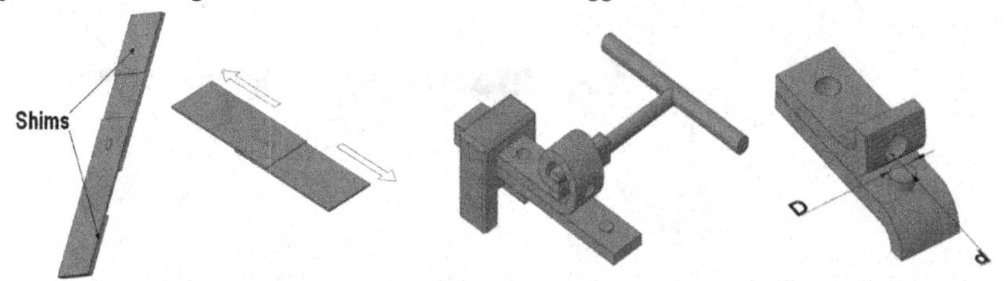

Figure 3. Schematics of lap shear tensile tests; arrows show the loading direction.

Figure 4. Peel test method and peel test.

RESULTS AND DISCUSSION

The cross-section macrostructure of the RSW DP steel and detailed micrographs corresponding to the fusion zone (FZ), the base metal (BM), and the heat affected zone (HAZ) (further divided into: coarse grain (CG), fine grain (FG) intercritical region (IC) and subcritical (SC)) are depicted in Figure 5. The location of all mentioned regions is indicated by numbers in Figure 5.

Ferrite matrix (bright regions) along with dispersed islands of martensite (tan coloured regions) is clearly revealed in BM zone micrograph (Figure 5). Presence of ferrite seemed to decrease at the ICHAZ region and switched gradually to fully martensite at the FGHAZ and CGHAZ; however, grain growth of martensite is clearly observed at the CGHAZ. On the other hand, a broken morphology is observed at the SCHAZ which confirmed tempering of martensite. SEM analysis made at the SCHAZ (Figure 6) well confirmed tempered martensite in the DP980 steel.

Figure 7 shows various hardness indentations of welds located at the SCHAZ of the DP980. The path of indentations is located 100 μm far from Ac_1 (line of lower critical temperature) as illustrated in Figure 7. As benchmark, the average for hardness in base metal is 340 HV.

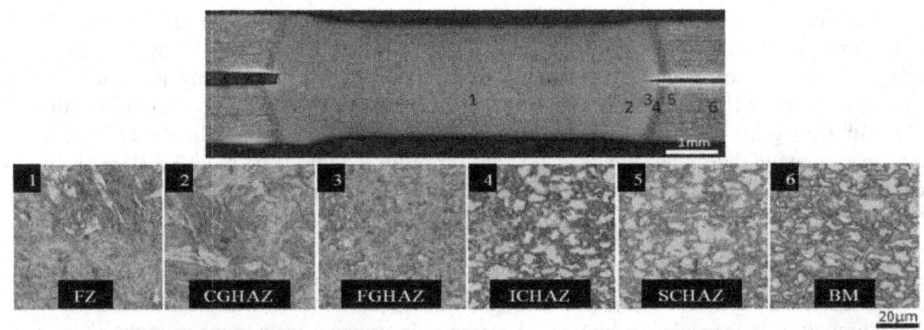

Figure 5. Macrographs of weld cross-sections and optical images taken from different regions in resistance spot-welded sample.

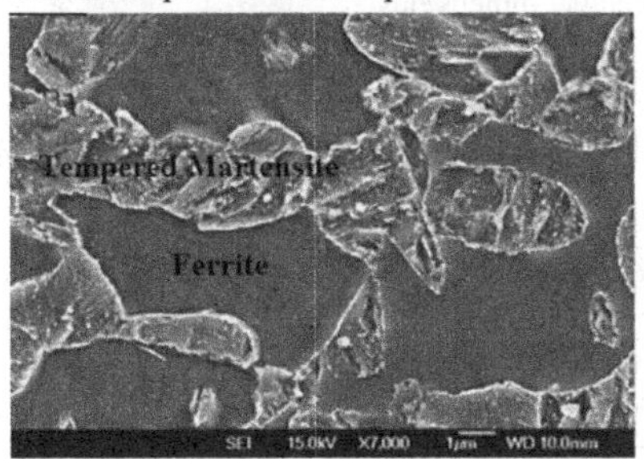

Figure 6. SEM images of SC-HAZ in DP980 steel.

Figure 7. Hardness profile of spot welds in DP980 steel.

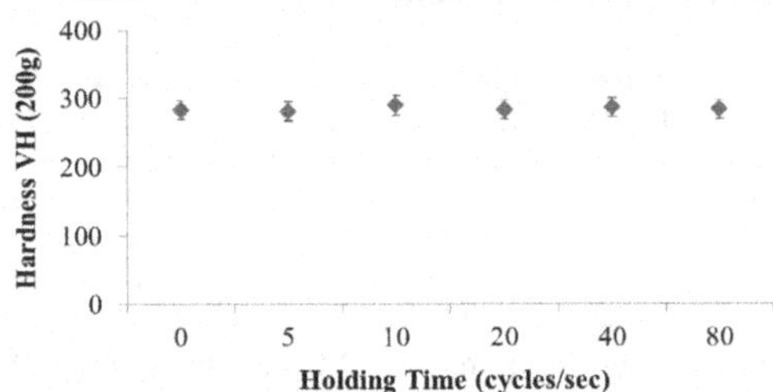

Figure 8. Hardness values obtained at the sub-critical HAZ of DP980 steel for different conditions of holding time.

Hardness measurements obtained at the SCHAZ of various specimens of holding time are provided in Figure 8. It is clear to note that a reduction in hardness (softening) occurred with respect to BM hardness (i.e. 340 HV) in all specimens; however, the influence of cooling rate (holding time) did not revealed significant differences. Furthermore, softening measured in SCHAZ is clearly associated to the tempering of martensite (Figure 6), and all assessed specimens revealed similar characteristics in terms of microstructure (i.e. broken morphology at the tempered martensite phase).

The average normalized peak loads to failure for all different conditions of holding time are given in Figure 9. As can be seen, the failure to load is quite consistent for all specimens and no major chances are observed. Cooling rate (holding time) had no effect on the mechanical performance of RSW DP980 steel.

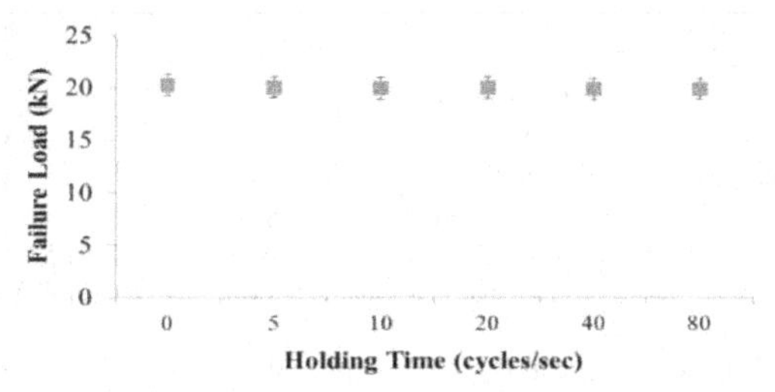

Figure 9. Peak load versus different conditions of holding time in DP980 welds

Figure 10 provides representative failure modes in all the samples. Pull-out failure mode is consistent in all the samples tested for each condition of holding time.

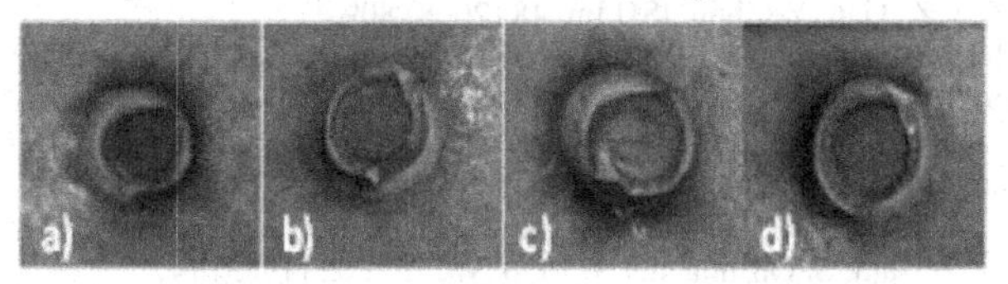

Figure10. Representative failure modes for (0, 5, 10 and 20 cycles) of holding time respectively

CONCLUSIONS

1. Insignificant changes in hardness is revealed at the sub-critical heat affected zone upon different conditions of holding time.
2. The cooling rate (holding time) range employed in this work did not show any effect on the tempering process of martensite phase in DP980.
3. According to the mechanical testing (lap-shear tensile test); softening (tempering of martensite) is barely affected by holding time (cooling rate) in resistance spot welded DP980 steel.

ACKNOWLEDGMENTS

The authors acknowledge the financial support from the Emerging Leaders in the Americas Program (ELAP) of International Scholarships Government of Canada. The authors also acknowledge the support of Arcelor Mittal Dofasco Inc. at Hamilton, Canada, HUYS Industries at Toronto, Canada, and the Autonomous University of Zacatecas, Zacatecas, México in carrying out this work. Authors are also thankful to MASc. Dave Huang, and Dr. S.S. Nayak of Center for Advanced Materials Joining at University of Waterloo, and Elliot Biro of Arcelor Mittal Global Research at Hamilton, Canada.

REFERENCES

1. Committee on Automotive Applications: “Advanced High Strength Steel (AHSS) Application Guidelines”, International Iron and Steel Institute (eds.), Version 4.1, June, pp. 1-4.
2. P. K. Ghosh, P. C. Gupta, R. Avtar , B. K. Jha (1990) ISIJ Int. 30 (1990) 233.

3. W. Tong, H. Tao, X. Jiang, N. Zhang, M. P. Marya, L. G. Hector, X. Q. Gayden. Metall. Mater. Trans. 36A (2005) 2651.
4. V. H. Baltazar Hernandez, Effect of Martensite Tempering on HAZ-softening and Tensile Properties of Resistance Spot Welded Dual-phase Steels, Ph.D. Thesis, University of Waterloo (Waterloo, 2010).
5. M. Marya, K. Wang, L.G. Hector, X. Gayden. J. Manuf. Sci. Eng. 128 (2006) 287.
6. S.K. Panda, M.L. Kuntz, Y. Zhou, Sci. Technol. Weld. Joining 14 (2009) 52.
7. V. H. Baltazar Hernandez, M. L. Kuntz, M. I. Khan, Y. Zhou. Sci. Technol. Weld. Join. 13 (2008) 769.
8. M. I. Khan, M. L. Kuntz, Y. Zhou. (2008) Sci. Technol. Weld. Join. 13 (2008) 49.
9. M. Xia, E. Biro, Z. Tian, Y. Zhou. ISIJ Intern., 48 (2008) 809.
10. M. I. Khan, M. L. Kuntz, E. Biro, Y. Zhou. (2008) Mater. Trans. JIM 49 (2008)1629.
11. S. K. Panda, N. Sreenivasan, M. L. Kuntz, Y. Zhou. (2008) J. Eng. Mater. Technol. 130 (2008) 041003.
12. M. Xia, E. Biro, Z. Tian, Y. Zhou. ISIJ Int. 48 (2008) 809.
13. G. Krauss. Steels, heat treatment and processing principles. ASM International, Materials Park, OH (1990).
14. D. Venugopalan, J. S. Kirkaldy. Hardenability concepts with applications to steel. In: Proceedings of a symposium held at Chicago, (1977) pp 249–272.
15. V.H. Baltazar-Hernandez, S.S. Nayak, Y. Zhou, Met. Mat. Trans. *A* 42, (2011) 3115.
16. N. Yurioka, H. Suzuki, S. Ohshita and S. Satio. Weld J., 62 (1983) 147.
17. RWMA: 'Resistance welding manual' 4th ed; 2003, Philadelphia, PA, RWMA, 18-1-18-14.
18. Recommended practices for test methods for evaluating the resistance spot welding behavior of automotive steels, ANSI/AWS/SAE/D8. 9-97, AWS, Miami, FL, USA, 1997.

Mater. Res. Soc. Symp. Proc. Vol. 1485 © 2013 Materials Research Society
DOI: 10.1557/opl.2013.256

Effect of Slag on Mixing Time in Gas-Stirred Ladles Assisted with a Physical Model

Adrián M. Amaro-Villeda[1], A. Conejo[2] and Marco A. Ramírez-Argáez[1].
[1] Facultad de Química, Universidad Nacional Autónoma de México, Mexico City, Mexico
adrianvilleda@yahoo.com.mx, marco.ramirez@unam.mx
[2] Centro de Graduados, Instituto Tecnológico de Morelia, Avenida Tecnológico 1500, Col Lomas de Santiaguito, Morelia, Michoacán, México.

ABSTRACT

A 1/6th water physical model of a 140 tons gas-stirred steel ladle is used to evaluate mixing times (τ_m at 95% of chemical uniformity) in a two phase system without slag (air-water) and in a more realistic three phase system (air-water-oil) to simulate the argon-steel-slag system and quantify the effect of the slag layer on the mixing time. Slag layer is kept constant at 0.004 m. Mixing times are estimated through measured changes in pH due to the addition of a tracer (NaOH 1 M). The effect of the following variables on the mixing time is evaluated for a single injector: gas flow rate (7, 17 y 37 *l*/min) and the injector position (R/r= 0, 1/3, ½, 2/3 and 4/5). Experimental results obtained in this work show good agreement when compared against mixing time correlations reported by Mazumdar for the two phase air-water case (no slag considered). Another comparison is done using the new concept called "effective bath height" proposed by Barati, where the mixing time is a function of the size of the slag layer since this layer dissipates part of the total amount of stirring energy introduced into the ladle by the injection of gas. Agreement is not good in this case. Finally, an estimation of the percentage of the stirring energy dissipated by the slag is computed, including other factors that govern the dissipation of stirring energy. Percentage of energy dissipated by the slag is found to be between 2.7 to 12 % depending on the process conditions.

INTRODUCTION

Many research works have been developed on the gas-stirred ladle furnace through physical and mathematical models as presented by the excellent reviews by Mazumdar [1,2]. Most of the models on ladle furnace through physical and/or mathematical models are carried out neglecting the presence of the slag layer. The slag phase plays several roles in the secondary steelmaking: a) to avoid reoxidation of liquid steel, b) to eliminate non-metallic inclusions, c) to eliminate sulphur, d) to prevent radiation losses from the steel to the surroundings.

There are only a few works about physical models of the ladle furnace that considered the three phase system gas-steel-slag [3-6]. These works evaluate several issues, being one of the most important to quantify how much is the mixing time increased by the presence of the slag layer. Most of the physical models considering the slag are based on a water-oil system. A great variety of oils have been used as well as other volatile and dangerous organic compounds to simulate the slag. However, usually these substances do not satisfy kinematic similarities or density ratio (steel density/slag density). Despite the similarity criteria are missed, contributions have been presented on the determination of the exposed areas during bottom blowing. One of the most important issues still to be defined is the effect of the presence of a slag layer on the mixing time. A few studies [2,3,8] report that mixing times increase when the slag is present.

However, the fraction of the stirring energy input that is dissipated by the slag phase has not been properly estimated yet. New ultrasonic techniques have been introduced in the literature to determine wave propagation in the free surface and the amount of energy consumed there, finding that this lost represent less than 1% at low flow rates [9]. The objective of this work is to determine the fraction of the stirring energy input dissipated by the presence of the slag phase with a 0.004 m of thickness and also to determine mixing times under the presence of slag assisted by a physical model.

EXPERIMENT

A transparent acrylic cylinder with a geometric scale factor λ=1/6 is used to simulate a 140 tons industrial ladle furnace. The physical model of a 0.52 m diameter and a 0.42 m of bath height operates at gas flow rates of 7, 17 and 37 l/min, injected through nozzles located at the bottom with different radial positions of r/R= 0, 1/3, ½, 2/3 and 4/5. Mixing times are determined by using the pH measurement technique and using the criteria of 95% of uniformity. A NaOH 1 M solution is used as a tracer to be added in the ladle. The Engine Oil with the properties presented in the Table I is selected after a screening process with several substances, since this oil satisfies better some of the similarity criteria. Physical properties of all the other substances in the model and in the ladle furnace are also presented in Table I. Further details on the experimental conditions and experimental setup can be found in [10].

Table I. Physical properties of the fluids in the physical and industrial systems

	Density (ρ, Kg/m^3)	Molecular viscosity (η, N s^2/m^2)	Kinematic Viscosity (ν, m^2/s)
Steel at 1600 °C	7000[11]	0.0068[12]	0.97X10^{-6}
Water at 20 °C	998.2[10]	0.0010[10]	1X10^{-6}
Slag	3000	0.18 to 0.35	6X10^{-5} to 12X10^{-5}
Engine Oil	890	0.19135	2.15X10^{-4}

RESULTS AND DISCUSSION

Figure 1 presents mixing times as a function of plug position for different gas flow rates and with or without slag. From this plot it is clear that regardless the gas flow rate, mixing time decreases when the plug is placed away from the center, reaches a minimum at position of r/R=1/3 and r/R=1/2, but increases again as the plug approaches to the ladle wall. The same trend is maintained when a 0.004 m oil layer is placed but mixing times are around 10 s longer than those mixing times obtained without slag.

Model to determine mixing times and stirring energies without slag

Equation (1) is used to predict mixing times without slag [1]. This equation is derived from a physical model study with a single plug located at the center of the bottom of the ladle (r/R=0) and varying the height of the bath (L) from 0.05 m to 1.1 m and maintaining a diameter of the ladle in 1.12 m (radius of ladle R=0.56 m). Stirring energy is determined by using equation (2) [8]. Equations (1) and (2) are derived by Mazumdar to predict mixing times as a function of the stirring energy and we used them to compare these predictions against our experimental measured mixing times (τ_m). Such a comparison is presented in Figure 2, where the mixing time is plotted as a function of the stirring energy (ε_m) for all cases with a single plug.

$$\tau_m = 37\varepsilon_m^{-0.33} L^{-1} R^{1.66} \quad (1)$$

$$\varepsilon_m = \frac{\rho_L g Q L}{\rho_L \pi R^2 L} \quad (2)$$

where ρ_L is the density of the water, Q is the gas flow rate and g is the gravity constant.

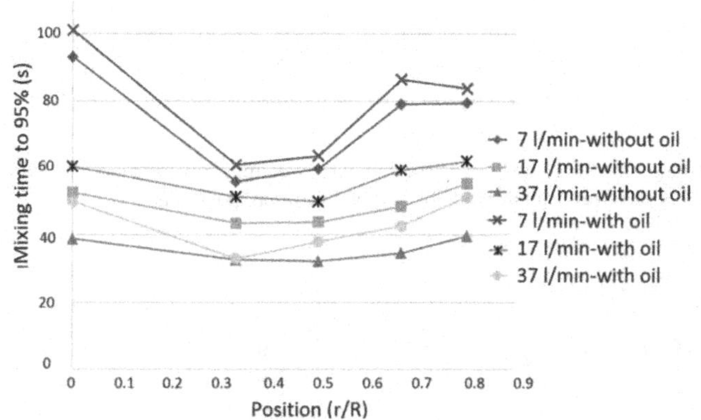

Figure 1: Mixing times as a function of plug position for a single plug. Every line represents a different gas flow rate and the presence or the absence of a slag layer.

From Figure 2 it is observed that experimental curves follow the same trend, i.e. the mixing time decreases as the stirring energy (ε_m) increases. The curve representing the correlation of Mazumdar (equation (1)) lies below our experimental data, and actually only our experimental curves obtained with plugs located at r/R= 1/3 and ½ are close to predictions from equation (2). The rest of experimental curves (with plugs at r/R=0, 2/3 and 4/5) show discrepancies with Mazumdar's correlation.

Mixing times estimated with the "effective height" parameter

Nowadays there are some efforts to express mathematically the effect of the presence of a slag layer on the top of the ladle furnace. Barati [7] introduced a new concept called "effective height", h_{eff}, which is represented by equation (3) as a function of the thickness of slag layer (h_s) and the height of liquid steel (h_m). Equation (4) is used to compute a blending energy ($\varepsilon_{blending}$) based on the stirring energy (ε_m) and by combining equation (3) with equation (4), mixing time (τ_m) can be predicted with equation (5).

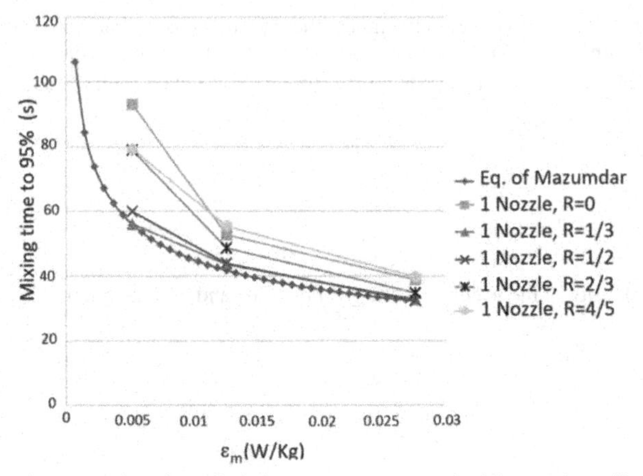

Figure 2: Mixing times as a function of stirring energy computed with equations (1) and (2) and measured in this work for a single plug located at 5 different positions.

$$h_{eff} = h_m + \frac{\rho_s}{\rho_m} h_s \tag{3}$$

$$\begin{cases} \varepsilon_{blending} = 1.41\exp\left(-3.16\frac{h_s}{h_m}\right)\varepsilon_m & \frac{h_s}{h_m} > 0.1 \\ \varepsilon_{blending} = \varepsilon_m & \frac{h_s}{h_m} \le 0.1 \end{cases} \tag{4}$$

$$\tau_m = 2.33\varepsilon_{blending}^{-0.34} h_{eff}^{-1} \tag{5}$$

where ρ_m is the bath density and ρ_s is the slag density.

Figure 3 shows mixing times against stirring energy predicted by using equations (3), (4) and (5) together with the experimental results obtained in this work with a single plug located at different positions. Our results follow the same trend as the Barati's correlation, but our measurements show greater mixing times values than Barati's predictions. None of the 5 curves (corresponding to different radial positions of the plug) approaches to Barati's predictions. However, Barati's correlation erroneously predicts that mixing times are reduced with the presence of a slag layer, which is contradictory to our results and common sense.

Computation of the energy dissipated by the slag in the physical model

Mazumdar [8] proposes several equations to estimate the magnitude of the factors that contributes to the stirring energy being dissipated in the ladle. Some of these factors (which do not include the slag layer yet) may be estimated by experimental measurements or by using equations (6), (7) and (8) which are related to the dissipation of energy due to the conversion of input energy to turbulence energy dissipation (η_{TED}), due to the bubbles (η_{Bub}) and due the walls (η_{Wall}) respectively. It is also assumed that when the opening of the slag occurs, during bubbling, the exposed surface present some factors of energy dissipated computed by equations (9) and (10). Calculations of all these factors are shown in Table II, which are used to estimate the effect of the slag layer on the mixing time in the ladle furnace. Operational conditions include a 0.004 m slag thickness, a plug position of r/R=4/5. When the gas flow rate is 7 l/min, the percentage of energy dissipated by the slag phase is estimated with equation (11) and it is found to be in the range of 4.5 to 12%. With a gas flow rate of 17 l/min, this percentage remains almost unchanged in values from 4.37 to 11.9% and for high gas flow rates of 37 l/min, those percentages decrease to a range between 2.68 and 10.2%.

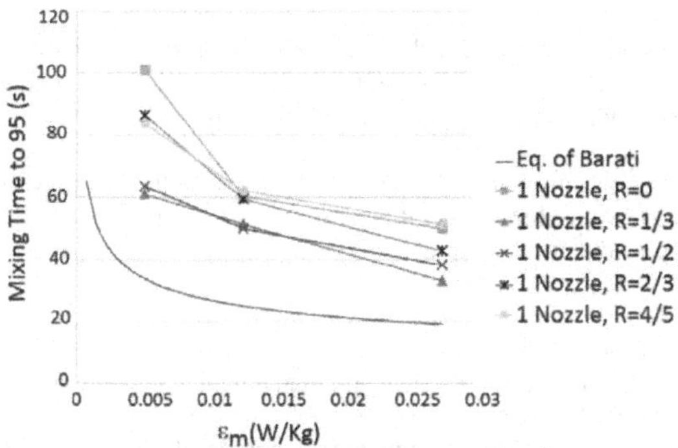

Figure 3: Mixing times as a function of the stirring energy using the concept from Barati [7] of "effective height".

Table II. Computed factors of dissipation of energy for all experimental conditions.

Without oil, r/R=4/5	Bubble velocity U (m/s)	η_{TED}	η_{Bub}	η_{Wall}		
1 nozzle, Q=7 l/min	0.05508	0.1060	0.341	0.445		
1 nozzle, Q=17 l/min	0.08324	0.1699	0.316	0.414		
1 nozzle, Q=37 l/min	0.12487	0.3059	0.294	0.322		
With oil, r/R=4/5	U (m/s)	η_{TED}	η_{Bub}	η_{Wall}	$\eta_{oil(maximum)}$	$\eta_{oil(minimum)}$
1 nozzle, Q=7 l/min	0.04407	0.0430	0.341	0.496	0.120	0.0451
1 nozzle, Q=17 l/min	0.06812	0.0755	0.315	0.491	0.119	0.0437
1 nozzle, Q=37 l/min	0.11017	0.1837	0.294	0.421	0.102	0.0268

$$\eta_{TED} = 0.618U^4R^{2.33}Q^{-1.33}L^{-1} \quad (6)$$

$$\eta_{Bub} = \frac{1.05Q^{0.2}}{3.8Q^{0.33}L^{0.25}R^{0.58} + 1.05Q^{0.2}} \quad (7)$$

$$\eta_{wall} = \frac{0.5R + L}{R + L}\left(1 - \eta_{TED} - \eta_{Bub}\right) \quad (8)$$

$$\left(\eta_{Free-Surface}\right)_{oil} = 0.0 \quad (9)$$

$$\left(\eta_{Free-Surface}\right)_{oil} \approx \left(\eta_{Free-Surface}\right)_{air} \approx 0.075 \tag{10}$$

$$\eta_{oil} = 1 - \left(\eta_{TEDoil} + \eta_{Buboil} + \eta_{Vessel-Wall} + \left(\eta_{Free-Surface}\right)_{oil}\right) \tag{11}$$

CONCLUSIONS

Experimental results of mixing times without slag with a single plug located at r/R=1/3 and ½ correlates well with empirical correlation proposed in literature (Equation 1) for all gas flow rates explored in this work.

Experimental results do not agree with the correlation reported by Barati. It is concluded that Barati's correlation erroneously predicts that mixing times are reduced with the presence of a slag layer, which is contradictory to ours and other results reported.

By using Mazumdar's correlation (Equation 11) it is found that the percentage of the stirring energy dissipated by the slag layer (oil layer of 0.004 m) is between 2.68% to 12 % depending on the values of the process variables.

REFERENCES

1. D. Mazumdar and R. Guthrie, "The physical and mathematical modelling of gas stirred ladle Systems," ISIJ int. **35** No.1, 1 (1995).
2. D. Mazumdar and J. W. Evans: Macroscopic models for gas stirred ladles, ISIJ Int. **44** No.3, 447 (2004).
3. J. W. Han, S. H. Heo, D. H. Kam, B. D. You J. J. Pak and H. S. Song: Transient fluid flow phenomena in a gas stirred liquid bath with top oil layer-approach by numerical simulation and water model experiments, ISIJ Int., **41** No. 10, 1165 (2001).
4. K. Krishnapisharody and G. A. Irons, "Modeling of slag eye formation over a metal bath due to gas bubbling," Metall. Trans., **37B**, 763 (2006).
5. P. Dayal, K. Beskow, J. Björkvall and D. Sichen, "Study of slag/metal interface in ladle treatment," Ironmaking & Steelmaking, **33**, No.6, 454 (2006).
6. M. Thunman, S. Eckert, O. Hennig, J. Björkvall and D. Sichen, "Study on the formation of open-eye and slag entrainment in gas stirred ladle," Steel research int. **78** No. 12, 849 (2007).
7. L. T. Khajavi and M. Barati, "Liquid mixing in thick-slag-covered metallurgical baths-blending of bath," Metall. Trans., **41B**, 86 (2010).
8. D.Mazumdar and R. I. L. Guthrie, "Modeling Energy Dissipation in slag-covered steel baths in steelmaking ladles," Metall. Trans., **41B**, 976 (2010).
9. Y. Kishimoto, Y. Sheng, G. A. Irons and Jen-Shih Chang: "Energy dissipation distribution in gas-stirred liquids", ISIJ Int. **39** No.2, 113 (1999).
10. A. M. Amaro-Villeda, J. A. González, M. A. Ramírez-Argáez: "Experimental study in gas-stirred ladles with and without the slag phase trough a water physical model," Mater. Res. Soc. Symp. Proc. Vol. **1373**, 155 (2012).

Mater. Res. Soc. Symp. Proc. Vol. 1485 © 2013 Materials Research Society
DOI: 10.1557/opl.2013.277

Synthesis and characterization of ceramic composites of the binary system $Ba_{0.75}Sr_{0.25}AlSi_2O_8$ - Al_2O_3

Jorge López-Cuevas, Magaly V. Ramos-Ramírez and José L. Rodríguez-Galicia
CINVESTAV-IPN, Unidad Saltillo, Carretera Saltillo Monterrey, Km. 13.5, C.P. 25900, Ramos Arizpe, Coahuila, México.

ABSTRACT

$Ba_{0.75}Sr_{0.25}AlSi_2O_8$ (SBAS) - Al_2O_3 composites, with SBAS/Al_2O_3 weight ratios of: (a) 90/10, (b) 70/30, and (c) 50/50, are *in situ* synthesized by reactive sintering at 900-1500°C/5h. The effect of mechanical activation of the precursor mixtures for 0, 4 or 8h in an attrition milling device on the microstructure and phase composition of the composites is studied. Only SBAS and Al_2O_3 phases are obtained at 1300-1500°C, independently of milling time. In general, the relative proportion of the desirable monoclinic SBAS (Celsian) phase increases in the materials with increasing milling time and sintering temperature, which is enhanced by their SrO content. The promotion of surface nucleation of the undesirable hexagonal SBAS (Hexacelsian) phase by mechanical activation results in a maximum Hexacelsian to Celsian conversion fraction of only 81.4%, obtained for composition 2 milled for 8h and sintered at 1500°C/5h. Under these synthesis conditions, an increment in the amount and size of the Al_2O_3 particles in the composites is detrimental for the Hexacelsian to Celsian conversion.

INTRODUCTION

Monoclinic $Ba_{1-x}Sr_xAl_2Si_2O_8$ (SBAS) solid solutions are interesting as matrixes for ceramic composites for structural applications at high temperatures, due to their high refractoriness, low coefficient of thermal expansion (CET), good resistance to oxidation and to slag attack, as well as good thermal shock resistance [1,2]. However, hexagonal SBAS tends to appear prior to monoclinic SBAS, frequently remaining in a metastable state at low temperatures. Hexagonal SBAS is undesirable due to its relatively high CET and because it transforms into an orthorhombic form during cooling, which causes differential volume changes that lead to microcracking and weakening of the material [1-3]. The formation of SBAS, and its hexagonal to monoclinic conversion, can be promoted by partial substitution of BaO by SrO in $BaAl_2Si_2O_8$ (BAS) [1,2]. Mixtures of BAS and $SrAl_2Si_2O_8$ (SAS) could also be used for this [4]. Due to their similar properties, when both BAS and SBAS have monoclinic structure, they are indistinctly denominated as Celsian (or Monocelsian); similarly, the corresponding hexagonal forms are both known as Hexacelsian. The formation of SBAS obeys Vegard's law and it can occur over the entire concentration range ($0 \leq x \leq 1$) [2,5]. D. Long-González et al. [2] recommend using $0.25 \leq x \leq 0.375$ in order to obtain Celsian with optimum properties. There are reasons to believe that SBAS and Al_2O_3 are chemically compatible, in such a way that they are able to coexist at equilibrium in a wide range of temperatures [6-8]. However, this has not been studied in detail so far. In this work, we study the effect of mechanical activation of the precursor mixtures on the microstructure and phase composition of $Ba_{0.75}Sr_{0.25}Al_2Si_2O_8$ - Al_2O_3 composites.

EXPERIMENTAL DETAILS

The raw materials employed are: Coal fly ash (FA); Al_2O_3 (purity of 99.99 wt.% and particle size <1 μm, HPA-0.5, SASOL, EUA); $BaCO_3$ (purity of 99.43 wt.% and particle size of ~3 μm, Alkem, México), and $SrCO_3$ (purity of 97.83 wt.% and particle size of ~4 μm, Solvay, México). The FA is conditioned by reducing its content of iron oxides by a wet magnetic separation process, and its chemical composition is determined by semiquantitative short wavelength dispersive X-Ray Fluorescence (XRF) spectroscopy using a S4 PIONNER BRUKER apparatus. This is (wt.%): 64.48% SiO_2, 27.28% Al_2O_3, 2.33% Fe_2O_3, 2.44% CaO, 0.64% MgO, 0.62% TiO_2, 0.28% K_2O, 0.18% Na_2O, and 0.19% of other oxides (ZrO_2, MnO_2, SrO, PbO and P_2O_5). Thermogravimetry (TGA) reveals that it contains also 1.73% of free C. All the TGA analyses are carried out up to 1400°C, employing a PERKIN ELMER Pyris Diamond TG/DTA apparatus, platinum crucibles and heating rate of 10°C/min in air. The conditioned FA has a particle size of ~30 μm, as determined by laser diffraction using a Coulter LS-100 apparatus. X-Ray Diffraction (XRD) indicates that this material is mainly composed by α-quartz and Mullite ($Al_6Si_2O_{13}$), plus ~70 wt.% of a silicoaluminous glassy phase. All the XRD studies are carried out using a Philips X-PERT equipment and Cukα radiation. The three SBAS/Al_2O_3 weight ratios studied for the composites, hereinafter referred to as compositions 1, 2 and 3, respectively, are: 1) 90/10, 2) 70/30, and 3) 50/50. Stoichiometric precursor mixtures are milled for 0, 4 or 8h in a Teflon-lined closed chamber laboratory attrition mill operated at 1700 rpm, employing 8 mol.% MgO-partially-stabilized ZrO_2 balls as milling media, with a balls to load mass ratio of 5:1, and using ethylic alcohol as dispersion medium. All precursor mixtures are analyzed by XRD and TGA, and then uniaxially pressed into 1.5g-cylinders with a diameter of 1.2 cm and a height of 0.5 cm, applying a load of 4 Ton, and subsequently sintered for 5h at 900-1500°C, using a Thermolyne 46120-CM-33 high temperature electric furnace, with heating and cooling rates of 2 and 5°C/min, respectively. The phase composition is verified in all cases by means of XRD and SEM/EDS analysis using in the latter case a Philips XL30 ESEM apparatus, employing an acceleration voltage of 20-30 kV, and a working distance of 10 mm. Prior to this, the sintered cylinders are cross-sectioned and one part of them is mounted in cold-cure epoxy resin and ground using SiC papers with successive grit sizes from 80 grit to 2400 grit. This is followed by polishing to a mirror finish using diamond particles with successive sizes of 3, 1 and ¼ μm, and by graphite-coating using a JEOL JEE-400 vacuum evaporator. The remnant parts of the sintered cylinders are milled in a planetary mill using agate balls and containers, until a particle size of ~150 μm is obtained. These powders are used for XRD analysis for phase identification and to determine a semi-quantitative Hexacelsian to Celsian conversion fraction (%), employing for this the method described by Y.-P. Fu et al. [9].

DISCUSSION

Characterization of the mechanically activated precursor mixtures

After 8h of milling, the mean particle size of the precursor mixtures decreases from 21.6 to 6.3μm for composition 1, from 17.7 to 3.9μm for composition 2, and from 13.7 to 2.8μm for composition 3. The particle size of the precursor mixtures decreases also with increasing content of Al_2O_3, since this is the finest raw material used.

The obtained TGA curves, not shown, reveal that all samples start losing weight since the very beginning of the heating stage, which is probably due to the release of water adsorbed on the surface of the particles. In all cases, with increasing milling time the rate of weight loss increases, while the end temperature for the weight loss stage is shifted toward lower values and the total weight loss decreases. The latter phenomenon suggests the occurrence of a partial decomposition of the carbonates during the milling stage, whose extension is proportional to the duration of this operation. All these effects are more pronounced with increasing carbonate content in the precursor mixtures, which suggests that $BaCO_3$ and $SrCO_3$ are the most affected by mechanical activation among the different components present in the precursor mixtures. This is consistent with the known fact that mechanical activation of $BaCO_3$ and $SrCO_3$ decreases both the temperature and the activation energy required for their thermal decomposition, increasing at the same time their reactivity towards other compounds mixed with them [10]. After 8h of milling, the observed temperature shift is from ~1033°C to ~883°C (ΔT = 150°C) for composition 1, with a corresponding total weight loss decreasing from 11.1% to 9.1% (Δm = 2%) after the same milling time. For composition 2, the corresponding temperature shift occurs from ~1020°C to ~920°C (ΔT = 100°C), with a total weight loss decreasing from 9.4% to 7.6% (Δm = 1.8%). Lastly, for composition 3, a similar temperature shift takes place from ~1013°C to ~923°C (ΔT = 90°C), with a total weight loss of ~6.4% (Δm ~ 0%), independently of milling time. Since the observed weight losses are mostly caused by the release of CO_2 associated with the occurrence of the solid state reactions, the obtained results mean that these reactions take place at lower temperatures, and at faster rates, in the mechanically activated materials. This could be attributed to the high degree of disorder introduced into the crystalline structure of $BaCO_3$ and $SrCO_3$ by the high-energy milling, as well as to the high degree of homogeneity and small particle size achieved in the precursor mixtures during the milling process.

Phase evolution and microstructure of the sintered composites

According to Figure 1, in general Celsian, Hexacelsian and Al_2O_3 phases are formed at all sintering temperatures and milling times employed. In a few cases the presence of small amounts of SiO_2 and $BaAl_2O_4$ secondary phases is additionally detected at relatively low sintering temperatures, with the second phase being more persistent. However, only SBAS and Al_2O_3 phases are found in all materials sintered at 1300-1500°C.

Only in the case of compositions 1 and 2 sintered at 1500°C there is a slight increase in the relative proportion of Al_2O_3 with increasing milling time. In all other cases, this is relatively unaffected by milling. On the other hand, in general the relative proportion of Celsian increases with increasing milling time and sintering temperature, while Hexacelsian shows the opposite behavior. Full Hexacelsian to Celsian conversion is never attained, which could be due to the promotion of the surface nucleation of Hexacelsian caused by the milling of the precursor mixtures [11,12]. The highest conversion values are obtained with 8h of milling and sintering at 1500°C, which ranged from 64% to 81.4%. The maximum value corresponded to composition 2.

The amount of Hexacelsian decreases in the composite materials with increasing sintering temperature, for all milling times employed, probably due to the formation of a larger amount of Celsian and/or a higher Hexacelsian to Celsian conversion promoted at higher sintering temperatures by the SrO contained in the materials.

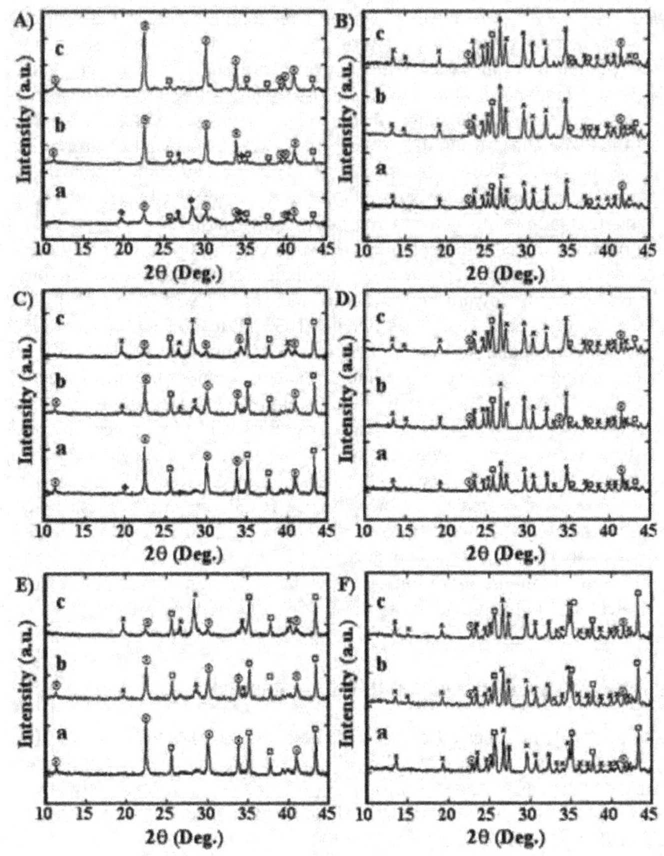

Figure 1. XRD patterns of ceramic composites synthesized at 900°C and 1500°C, of compositions 1 (A and B), 2 (C and D), and 3 (E and F), with milling times of: a) 0h, b) 4h and c) 8h. Key: •SiO_2 (Quartz), ❖$BaAl_2O_4$, □Al_2O_3 (Corundum), ×Celsian, and ⊙Hexacelsian.

All materials sintered at 900°C show generally a poorly-defined microstructure, Figure 2, which is probably due to an unfinished solid state reaction as well as to a lack of densification. In contrast, at 1500°C a well-defined microstructure constituted by two phases is observed, which is more evident for compositions 1 and 3. These phases are: 1) Dark grey platelet and blocky-shaped Al_2O_3 particles, and 2) light-colored SBAS (Hexacelsian + Celsian) polyhedral grains. The particle size of both phases ranged from ~5 μm to ~10 μm. In both compositions, a large number of intergranular microcracks is formed in the SBAS phase during cooling of the materials after the sintering stage, probably due to a mismatch existing in the thermal expansion coefficients of Hexacelsian, Celsian and Al_2O_3 phases.

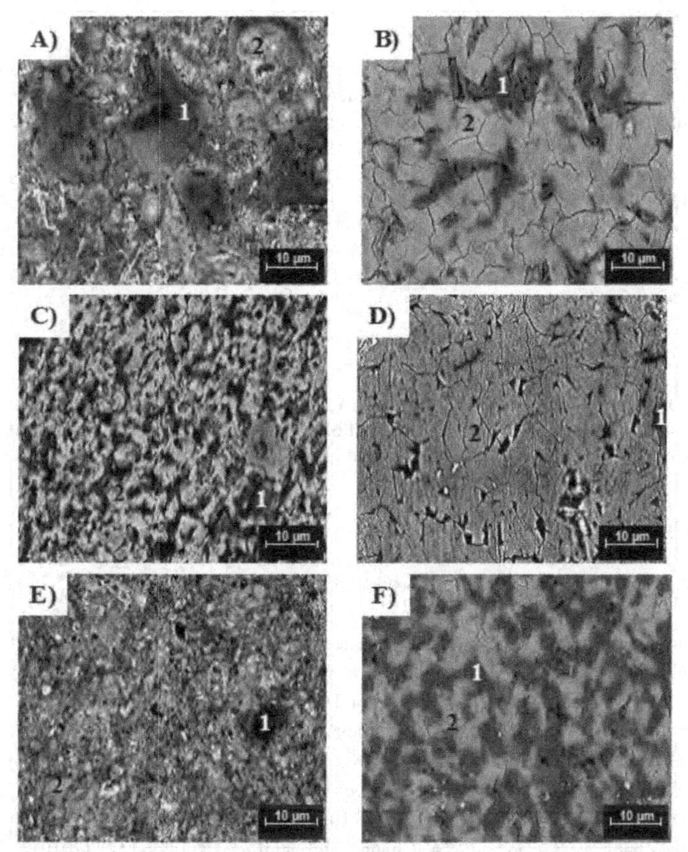

Figure 2. SEM micrographs of ceramic composites synthesized at 900°C and 1500°C, of compositions 1 (A and B, respectively), 2 (C and D, respectively), and 3 (E and F, respectively), in all cases with milling time of 8h.

For composition 2, the presence of the Al_2O_3 phase is not clearly evident due to its small crystal size, but the presence of light-colored SBAS polyhedral grains with a size of ~10 µm, as well as of intergranular microcracks, is clearly visible. In this case, the amount of porosity and the pore size are larger than in the other two compositions investigated. The larger amount and size of the Al_2O_3 particles seen in compositions 1 and 3 could further promote the surface nucleation of Hexacelsian, and this explains their lower Hexacelsian to Celsian conversion values. A similar situation has been reported for $Ba_{0.75}Sr_{0.25}Si_2Al_2O_8/Si_3N_4$ [13] and $Ba_{0.75}Sr_{0.25}Al_2Si_2O_8$/Mullite composites [14].

CONCLUSIONS

The solid state reactions take place at lower temperatures and at faster rates in the milled materials, with a larger effect produced with increasing carbonate content in the precursor mixtures. Only SBAS and Al_2O_3 phases are obtained at 1300-1500°C/5h, independently of milling time. Since mechanical activation promotes the surface nucleation of Hexacelsian, it obstructs the attainment of full Hexacelsian to Celsian conversion in the synthesized composites. In spite of that, in general the relative proportion of Celsian increases with increasing milling time and sintering temperature, which could be enhanced by the SrO contained in the materials. For 8h of milling and sintering at 1500°C/5h, an increased amount and size of the Al_2O_3 particles in the composites is detrimental for the Hexacelsian to Celsian conversion.

ACKNOWLEDGMENTS

The authors express their gratitude to CONACYT and Cinvestav-Saltillo for the financial support and facilities provided for the development of this work, as well as to the personnel of the "José López Portillo" power plant, Nava, Coahuila, México, for supplying the employed FA.

REFERENCES

1. N.P. Bansal, M.J. Hyatt and C.H. Drummond III, *Ceram. Eng. Sci. Proc.* **12**, 1222-1234 (1991).
2. D. Long-González, J. López-Cuevas, C.A. Gutiérrez-Chavarría, P. Pena, C. Baudin and X. Turrillas, *Ceram. Int.* **36**, 661-672 (2010).
3. H.C. Lin and W.R. Foster, *Am. Mineral.* **53**, 134-144 (1968).
4. N.A. Sirazhiddinov, P.A. Arifov and R.G. Grebenshchikov, *Inorg. Mater.* **8**, 756 (1972).
5. N.P. Bansal, *J. Mater. Sci.* **33**, 4711-4715 (1998).
6. C.E. Semler and W.R. Foster, *J. Am. Ceram. Soc.* **52**, 679-680 (1969).
7. P.S. Dear, *Bull. Virgina Polytechnic Inst.* **50**, 8 (1957).
8. C. Zhang, F. Zhang, W.S. Cao and Y.A. Chang, *Intermetallics* **18**, 1419-1427 (2010).
9. Y.-P. Fu, C.-C. Chang, C.-H. Lin and T.-S. Chin, *Ceram. Int.* **30**, 41-45 (2004).
10. J.M. Criado, M.J. Diánez and J. Morales, *J. Mater. Sci.* **39**, 5189-5193 (2004).
11. S. Bošković, D. Kosanović, Dj. Bahloul-Hourlier, P. Thomas and S.J. Kiss, *J. Alloy. Compd.* **290**, 230-235 (1999).
12. S. Bošković, Dj. Kosanović and S. Zec, *Powder Technol.* **120**, 194-198 (2001).
13. L. Limeng, Y. Feng, Z. Haijiao, Y. Jie and Z. Zhiguo, *Scripta Mater.* **60**, 463-466 (2009).
14. J. López-Cuevas, D. Long-González and C.A. Gutiérrez-Chavarría in *Advanced Structural Materials-2011*, edited by H.A. Calderon, A. Salinas-Rodriguez and H. Balmori-Ramirez, (Mater. Res. Soc. Symp. Proc. **1373**, Cambridge University Press, N.Y., 2012) pp. 43-52.

Mater. Res. Soc. Symp. Proc. Vol. 1485 © 2013 Materials Research Society
DOI: 10.1557/opl.2013.278

Effect of the ratio Mo/Cr in the precipitation and distribution of carbides in alloyed nodular iron

H.D. Rivero[1], José .A. García[1], E. Cándido Atlatenco[1], Alejandro D. Basso[2], J. Sicora[2]

[1]Departamento de Ingeniería Metalúrgica, Facultad de Química, Universidad Nacional Autónoma de México.
[2]INTEMA-Universidad Nacional de Mar del Plata, Argentina

ABSTRACT

This investigation deals with the effect of 2/0, 1/1 and 1/0.5 Cr/Mo ratios on the local fraction, distribution and the comparative size of carbides precipitated in cast nodular iron. "Y" block castings with a thickness of 1.5 cm are cast in green sand molds. Two samples are cut from each casting, one located on the center and another on the wall. The carbide volume fraction is evaluated by a digital analysis system. Each sample is analyzed in three zones: bottom, middle and top. Carbide mappings are generated according to the average local carbide fraction in order to get the distribution of carbides on the casting. Results show that higher volume fractions of carbides precipitate for the ratio 2/0 of Cr/Mo with values between 28.5 and 19.5%. The lowest fraction of carbides is presented in nodular iron alloyed with a Cr/Mo ratio of 1/1 between 6.5 and 4.6%. Also a very heterogeneous distribution of the carbides is observed in the three alloys and massive carbides are observed in the last freezing zone of the castings.

INTRODUCTION

The manufacturing of conventional nodular iron with carbides is a topic associated to its negative effect on the mechanical properties. However recently it has been more intensely investigated because Carbide Ductile Iron (CDI) is important for the manufacturing of Carbide Austempering Ductile Iron (CADI) in components with high wear and abrasion resistance. These characteristics are associated to the residual fraction of carbides in the ausferrite matrix. In CADI materials the carbides work as reinforcements of the matrix. According to its microstructure and properties CADI materials have important applications in mining, construction, agricultural and other industries. CDI can be produced by casting processes controlling the chemical composition (alloying elements or C-Si balance) and cooling rate during solidification. Some alloying reduces the difference between the eutectic temperature of the Fe-C stable and metastable diagrams [1] that promote the precipitation of carbides and graphite. Whitening elements such as Cr, V, Mo and B are strong carbide promoters in nodular cast irons, however they can cause a heterogeneous distribution of carbides and strong segregation patterns in the Last Freezing Zones (LFZ) [2, 3]. The aim of this work is to determine the effect of three different Cr/Mo ratios on the average local fraction carbide, carbide mapping (distribution), total carbides % in sample located in the wall with a high cooling rate and in the center with a slow cooling rate.

EXPERIMENT

Nodular irons whit three Cr/Mo ratios i.e., 2/0, 1/1 and 1|/0.5, are manufactured in an induction furnace with a 120 Kg of capacity. Raw materials are high manganese steel scrap and scrap of cast iron. The chemical composition adjustments are made with FeSi75, FeMn70, pure Cr, FeMo74 and pure copper, graphite is used as a recarburant agent. The inoculation is performed with FeSi75 and a nodulizing treatment with FeSi45Mg6. Y block castings of 1.5 cm in thickness are poured in green sand molds. Two samples of each casting located one on the wall and another in the center are cut out, Figure 1a. Each sample is analyzed in three positions: wall, intermediate and center (horizontal direction) and also three positions in the vertical direction: bottom, center and top, Figure 1b. At each position are taken three images to get the average local carbides % whit the purpose to get the distribution of carbides in zones with high cooling rate (wall sample) and slow cooling rate (center sample) in the same casting. Samples are etched with a fresh solution of HNO_3 10 mL, HF 4 mL and 87 mL H_2O. The fraction of carbides is performed by image processing using Image Pro Plus software at a 100x magnification. Carbide distribution mappings are obtained for the two samples and each casting.

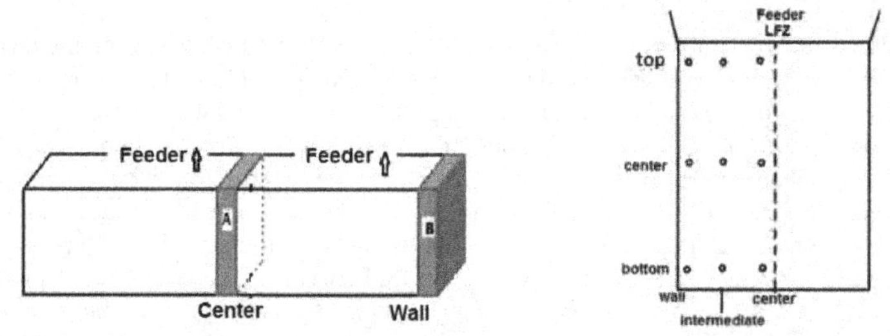

Figure 1. (a) Samples cut from the Y block casting: center and wall. (b) Local zones analyzed in each sample from the bottom to top (vertical analysis) and wall to center (horizontal analysis). LZF is the Last Freezing Zone.

DISCUSSION

The characteristics of the graphite for the three experimental nodular irons are shown in Table I. The matrix in all cases is perlite and carbides. Non etched microstructures of the three nodular iron are shown in figure 2. Nodular iron with ratio 2/0 has lower nodule count and lower nodularity %. This is due to the larger fraction of carbides that precipitate and reduce the amount of available carbon available for graphite nodules. Some nodules lose their morphology and precipitate together with carbides in the matrix. The results of the carbide mapping and the carbide local average fraction (%) in the horizontal (wall to center) and vertical (bottom to top) directions for both of them samples wall and center are showed in figures 3a and 3b for Cr/Mo = 2/0, figures 4a and 4b for Cr/Mo = 1/1 and figures 5a and 5b for Cr/Mo = 1/0.5.

Table I. Characteristics of graphite nodules as a function of the Cr/Mo ratio

	AFS Characteristic [4]		
Cr/Mo Ratio	Nodularity %	Nodule Size	Nodule Count(AFS)
2/0	90	8-9	200
1/1	95	7-8	200-250
1/0.5	95	7-8	200-250

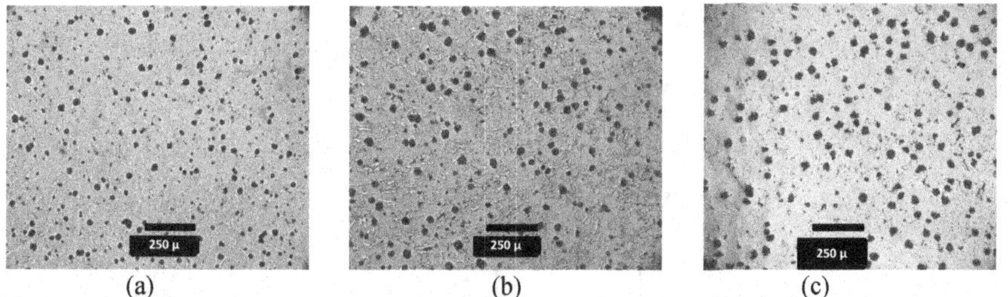

Figure 2. Microstructure showing graphite nodules (a) Cr/Mo = 2/0, (b) Cr/Mo = 1/1 and (c) Cr/Mo = 1/0.5. Non etched.

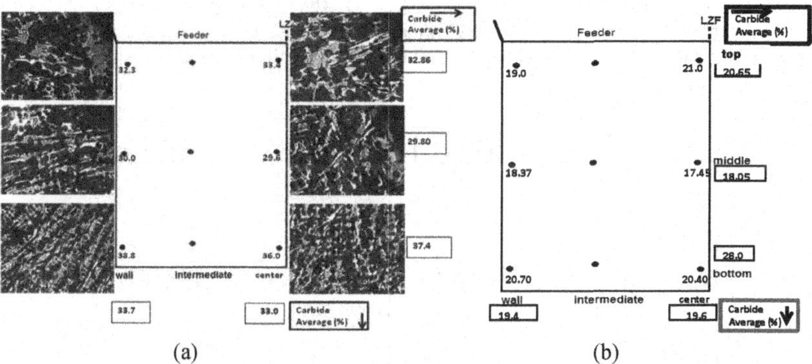

Figure 3. Average local fraction of carbides and average carbide % in the vertical and horizontal directions in the casting with ratio Cr/Mo = 2/0 (a) wall sample showing the carbide mapping microstructures and (b) Center sample just with average local carbide %.

Table II shows the comparative total average carbide % between the samples located at the wall and center for each Cr/Mo ratio, it also shows the variations between the wall and center samples.

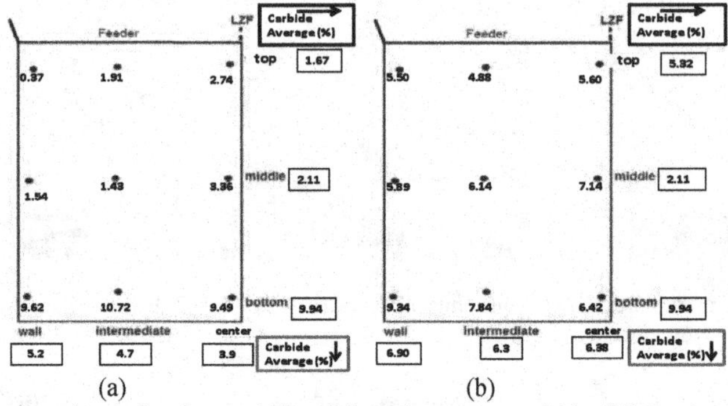

Figure 4. Average local fraction carbides (%) and average carbide (%) in the vertical and horizontal directions in the casting with ratio Cr/Mo = 1/1 (a) wall and (b) center sample

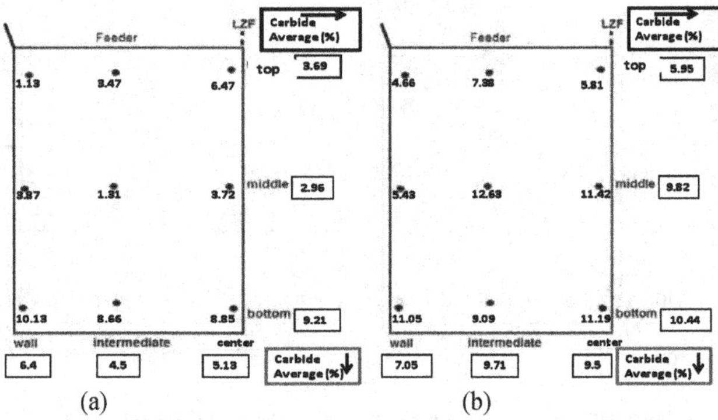

Figure 5. Average local fraction carbides (%) and average carbide (%) in the vertical and horizontal directions in the casting with ratio Cr/Mo = 1/0.5 (a) wall and (b) center sample

Table II. Total average carbides in the wall and center samples for each Cr/Mo ratios.

Cr/Mo ratio	Wall sample Carbide %	Center sample Carbide %	Difference	Variation %*
2/0	28.5	19.5	9.0	31.6
1/1	4.57	6.52	2.02	29.9
1/0.5	5.28	9.71	4.43	45.62

* Variation % =(higher %-smaller%)100/higher%

Table III. Difference between the maximum and minimum local average carbide % for the three Cr/Mo ratios as a function of the sample position (wall and center).

Cr/Mo ratio	Wall sample				Center sample			
	max	min	difference	% difference	max	min	difference	% difference
2/0	34.6	23.3	11.3	32.7	21.0	17.95	3.55	17.0
1/1	10.72	0.37	10.35	96.0	9.34	4.48	4.56	48.8
1/0.5	10.13	1.13	9.0	88.8	11.42	4.66	9.71	85.0

Table II shows the strong variations of carbide content between the samples located at the wall and in the center of the block and also between the Cr/Mo ratios. The variation of carbide contents between wall-center samples means that the local cooling rate has a strong effect on the size and fraction of precipitated carbides. However it is observed that the fraction of precipitated carbides does not follow the empirical rule that high cooling rates promote greater carbide formation. The finest carbides can be associated to a large undercooling in combination with the whitening effect of chrome. On the other hand, large carbides precipitate in the last freezing zone (LFZ) are associated to slow cooling rates with relative low undercooling values that promote few nucleating sites with favorable growing conditions. This last behavior can be explained accordingly with the segregation factor of chromium of 11.6, which promotes segregation of chromium in the last freezing zones and the formation of alloyed carbides [5]. Nodular iron with just 2.2 % Cr added as whitening element shows a high precipitation of carbides but with a very heterogeneous distribution and different sizes. When Cr and Mo are simultaneously added as whitening elements the effect on carbide formation is significantly reduced. The reason of this behavior can be explained by considering the effect of Cr and Mo on the graphite and white iron eutectic temperatures [2]. The Cr alloying reduces the temperature difference between the graphite and white iron eutectic temperatures promoting solidification below the Fe-carbide eutectic temperature, while Mo simultaneously depress both Fe-graphite and Fe-carbide eutectic temperature without reducing the difference between them. This promotes a solidification condition that follows more closely the Fe-graphite phase diagram. Such a competition between the effects produced by Cr and Mo reduces the precipitated carbide fraction when the Cr/Mo ratios are 1/1 and 1/0.5. Table III shows the difference between the maximum and minimum local average carbide % for the three Cr/Mo ratios studied as a function of the sample position (wall and center). This shows very high differences and its indicative of the heterogeneity of carbides in the casting. Both Cr or the combination Cr/Mo promote carbide precipitation in the nodular iron, but without ensuring a homogeneous carbide distribution.

CONCLUSIONS

The three studied Cr/Mo ratios in this work have no significant effect on % nodularity, size and nodule count. Nodular iron alloyed just with chromium presents only a slightly lower nodule count and size. The studied Cr/Mo ratios do not reduce the effect of the local cooling rate i.e., they do not follow the role of the cooling rate on the solidification of cast iron. The ratio Cr/Mo of 2.2/0 promotes the highest fraction of carbides between 28.5% and 19.5% with a strong variation % between the wall and the center samples of 31.6%. The ratio Cr/Mo of 1/1 presents the lowest carbide fraction between 6.52% and 4.57%, the % carbide variation from wall and center samples is 29.1%. The ratio Cr/Mo of 1/0.5 shows a carbide fraction between 9.71% and 5.28% with a % carbide variation of 45.6 %. The local solidification of the casting affects the carbide size precipitated during solidification.

ACKNOWLEDGMENTS

We thank PAPIIT IT118411-3 for the financial support granted as well as M. en I. Agustín G. Ruiz Tamayo for his support in the manufacture of nodular iron and IQM Victor A. Aranda Villada for his help in the metallographic evaluation of carbides.

REFERENCES

1. R. Biggs, Tramp elements in Ductile Iron, Iron Casting Research Institute, 2008
R. Bigge, Tramp elements in Ductile Iron, Iron Casting Research Institute 2009, p. 1-12, www.ductile.org/magazine/2009_2/Tramp elements
2. Z. Jiyang, Dalian University of Technology China, Serial Report February 2009
Z. Jiyang, Colour Metallography of Cast Iron, Dalian University of Technology China, Serial Report February 2009, http://www.foundryworld.com/uploadfile/200952136538017.pdf
3. S. Laino, R. Dommarco, J.A. Sikora, Desarrollo de fundiciones nodulares austemperadas con carburos CADI, Actas Congreso CONAMET-SAM , La Serena Chile, 2004. http://www.materiales-sam.org.ar/sitio/biblioteca/laserena/102.pdf
4. American Foundrymen´s Society (AFS), "Foundrymen´s guide to ductile iron microstructures" Des Plaines, Illinois, (1984).
5. J.D. Mullins, Basic Ductile Iron Alloyings, Metal Sorel Rev. March 2006 http://www.sorelmetal.com/en/publi/PDF/085_(2006).pdf

Mater. Res. Soc. Symp. Proc. Vol. 1485 © 2013 Materials Research Society
DOI: 10.1557/opl.2013.279

Decarburization of Hot-Rolled Non-Oriented Electrical Steels

Emmanuel J. Gutiérrez[1] and Armando Salinas[2]
[1]Centro de Investigación y de Estudios Avanzados del Instituto Politécnico Nacional, Saltillo, Coahuila, PA 663, México

ABSTRACT

The high temperature decarburization-oxidation behavior of hot rolled, non-oriented electrical steel strips is investigated during air-annealing treatments. Annealing temperature and time are varied from 700 – 1050 °C and 10 to 150 min, respectively. The experimental results show that uniform external oxidation affects strongly the rate at which carbon can be removed from this material. The thickness of the oxide layer formed after 150 minutes of annealing increases linearly with increasing temperature in the range 828 and 920 °C. The effect of temperature on the thickness of the oxide scale at temperatures outside this range is significantly smaller. These results indicate that the rate of oxidation in this material is strongly influenced by the microstructure of the steel during annealing. Decarburization rates are very slow during annealing at $T \leq 750$ °C where the oxide layer is thin and porous. In contrast, fast and intense decarburization of the strips is observed as a result of annealing at temperatures between 800 and 850 °C. Finally, decarburization at $T \geq 875$ °C becomes slower as the temperature is increased until at $T \geq 950$ °C this process is practically inhibited. Measurements of C content as a function of time and temperature show that the observed decarburization kinetics follows Wagner's model at 800 and 850 °C. However, at higher annealing temperatures decarburization is slower than that predicted by the model. This behavior is related to the increment of the oxide scale thickness and a transition from cracked to crack-free oxide structure which makes C diffusion through the oxide film very difficult.

INTRODUCTION

Electrical energy in terms of available resources, environmental and economic impact has become of great concern worldwide. Therefore, there is a driving force to produce more efficient electromagnetic components that allow minimizing the energy consumption. Electrical steels play an important role in the generation (generators), transmission and distribution (transformers) and consumption (motors) of electrical power and are the most important among the magnetic materials produced today [1]. Magnetic properties of these materials are affected by texture, grain size, second phase particles and residual stresses, which depend strongly on the processing variables [2]. Conventional manufacture of grain non-oriented (GNO) electrical steels involves several processing stages such as: hot rolling, cold rolling, intermediate annealing, temper-rolling and final decarburization annealing [2]. This last process is traditionally performed at temperatures below the Ac_1 transformation temperature (α-Fe→α-Fe+γ-Fe) during prolonged times (t □12 h), which implies high processing cost and consequently high cost of the final product. In a previous work [3], it has been demonstrated that from the microstructural, magnetic and mechanical properties point of view, annealing prior to cold rolling represents an attractive alternative method for the manufacture of low-C GNO steel sheets. In the present investigation, the effects of annealing time and temperature on decarburization of hot-rolled GNO electrical steels are investigated using temperatures within the single (α-Fe, γ-Fe) and two-phase (α-Fe+γ-Fe) fields and a maximum soaking time of 150 min.

EXPERIMENTAL DETAILS

Samples taken from a hot-rolled GNO electrical steel coil are subjected to annealing in still air at temperatures between 700 and 1050 °C in a muffle-type furnace (Thermolyne-6020). The heating rate used is 15 °C/min and soaking times are 30, 60, 90, 120 and 150 min followed by air cooling. The chemical composition of the as-received hot-rolled bands is presented in Table I. The C and S contents of the starting steel and annealed samples are determined in a LECO CS 230 carbon/sulfur determinator by combustion infrared absorption spectrometry based on ASTM E-1019. The Ac_1 and Ac_3 transformation temperatures are determined by differential thermal analysis (DTA) in a Perkin Elmer analyzer. Material under investigation, 10 mg of steel burr, is heated at rates of 10, 20, 30, 40 and 50 °C/min from room temperature up to 1150 °C. Duplicate experiments are performed under Ar atmosphere. Characterization of the oxide film by X-ray diffraction (XRD) is performed in an X′Pert–Philips diffractometer. Measurements are obtained in a 2θ range from 10 to 100° using CuK_α radiation with λ=1.5405 Å. Oxide films are also analyzed by scanning electron microscopy (SEM) in a Philips–XL30 ESEM microscope.

Table I. Chemical composition of GNO electrical steel [wt. %].

C	Si	Al	S	Mn	P	Cu	Cr	Ni	N_2
0.05	0.57	0.21	0.004	0.32	0.042	0.029	0.019	0.029	0.0056

RESULTS AND DISCUSSION

Fig. 1a shows the Ac_1 and Ac_3 transformation temperatures determined by DTA as a function of the heating rate. As can be observed, the transformation temperatures increase with increments in the heating rate. This is related to the reduction of available time for the phase transformation as the heating rate increases. For the heating rate used during annealing of the hot-rolled bands (15 °C/min), the Ac_1 and Ac_3 are 763 and 946 °C, respectively (see Fig. 1a). Fig. 1b illustrates the effect of time and annealing temperature on the C content of heat-treated samples. As can be seen, annealing at 700 °C did not cause decarburization of the steel independently of time (curve *A*). With increments in the annealing temperature up to 750, 800 and 850 °C, for a giving time, decarburization is more intensive as the temperature increases (curves *B, C* and *D*). However, when the temperature is raised up to 875 and 900 °C, decarburization becomes slower and the effect is greater at a higher temperature (curves *E* and *F*) until at temperatures of 950 and 1050 °C, decarburization is inhibited (curves *G* and *H*).

The kinetics of decarburization of Fe-C alloys with a microstructure consisting mainly of α-Fe+γ-Fe has been established by Wagner and developed with some simplified considerations by Swisher, Pyyry and Kettunen. These authors found that decarburization of steel can be described by the following equation [4]:

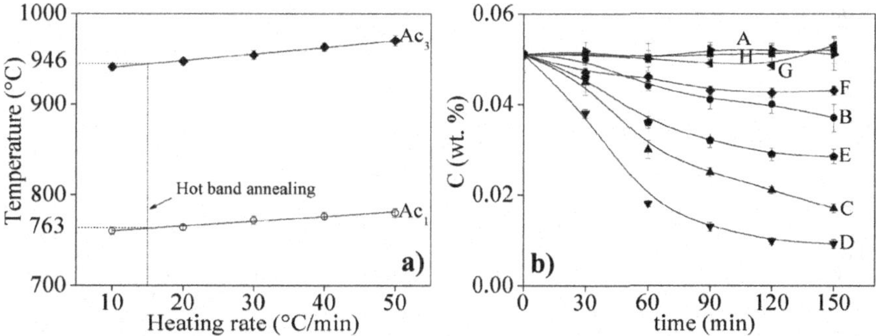

Figure 1. a) Ac_1 and Ac_3 transformation temperatures determined by DTA and b) evolution of C content during annealing of hot bands at: A) 700 °C, B) 750 °C,C) 800 °C, D) 850 °C,E) 875 °C, F) 900 °C, G) 950 °C, H) 1050 °C.

$$\overline{C} = C_i - \frac{1}{a}\sqrt{\frac{6D_f C_b t}{3C_i - bC_b}}\left(C_i - \frac{C_b}{2}\right) \tag{1}$$

where $\overline{C}$ is the average C concentration after decarburization, C_i is the initial carbon concentration, a is the steel half-thickness in cm, D_f is the diffusion coefficient of C in α-Fe in cm²/s, C_b is the carbon concentration in α-Fe in equilibrium with γ-Fe, t is time in s, and the constant b is 2 according to Swisher and 3 according to Pyyry and Kettunen (P-K). All carbon concentrations are in wt. %. According to Smith and Swisher, D_f and C_b can be calculated as follows [4]:

$$D_f = 0.256e^{-24,000/RT} \tag{2}$$

$$C_b = 0.1295 - 1.099\,x\,10^{-4}T \tag{3}$$

where R=1.986 cal/mol·K and T is temperature in K. Experimental C contents(C_{exp}) are compared with those determined theoretically by the kinetics model (C_{mod}) considering $b = 3$. Fig. 2 presents the results obtained for temperatures between 800 and 900 °C (two-phase field region according to results of Fig. 1a). The solid straight lines traced in these figures indicate C_{exp}=C_{mod}. As can be seen, when annealing is performed at 800 °C, C_{exp} values exhibit a good correlation with those predicted by the model (Fig. 2a). Annealing at 850 °C, results in an increase of the deviation between C_{exp} and C_{mod} values but there is still a good approximation (Fig. 2b). In contrast, at higher annealing temperatures, 875 and 900 °C, the experimental values describe a significantly slower decarburization than that predicted by the model even for short annealing times (Fig. 2c and 2d, respectively).

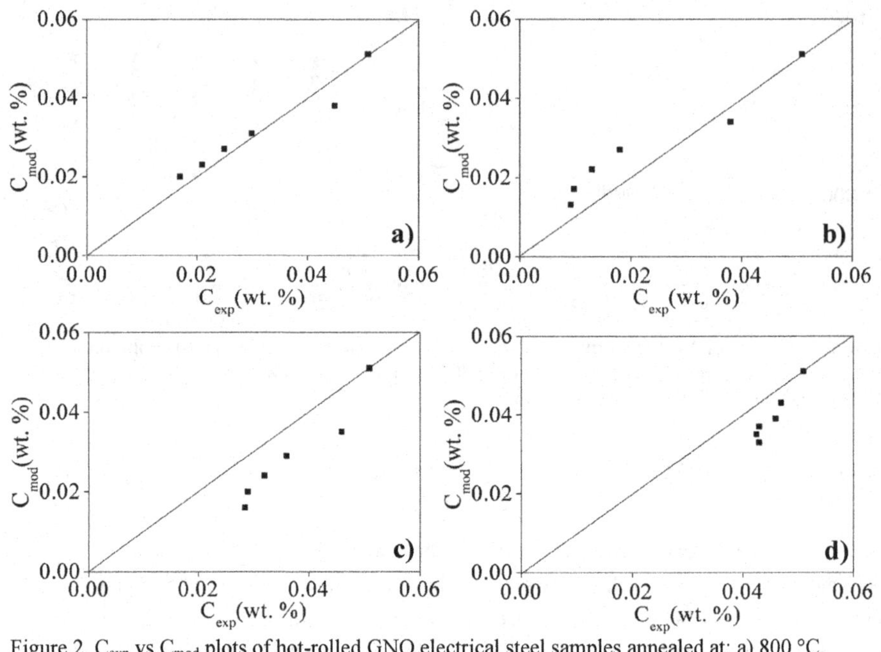

Figure 2. C_{exp} vs C_{mod} plots of hot-rolled GNO electrical steel samples annealed at: a) 800 °C, b) 850 °C, c) 875 °C and d) 900 °C.

Decarburization of steel during air-annealing treatments occurs simultaneously with oxidation according to the following reactions [5]:

$$[C] + FeO = Fe + CO \quad (4)$$

$$CO + FeO = Fe + CO_2 \quad (5)$$

Therefore, decarburization can only proceed when gases escape from the oxide scale. Due to this, oxide characteristics are very important to decarburization [5]. Fig. 3 shows the microstructural features of oxides formed during annealing of hot-rolled bands. Porous and cracked oxide structures are observed when annealing is performed at temperatures in the α-Fe phase field or at temperatures low within the two-phase field (T ≤ 850 °C) as shown in Figs. 3a and 3b, respectively. In contrast, at temperatures high within the intercritical region (T ≥ 875 °C) or in the γ-Fe single phase field, crack-free structures are observed (Fig. 3c).

The thickness of the oxide layer also changes significantly depending on the annealing temperature (Fig. 4a). Thin oxide layers are observed at temperatures under 800 °C. When the temperature is increased from 828 to 920 °C the oxidation rate increases linearly with temperature. The increment in the oxidation rate in the two phase field region has been associated by other authors to the α-Fe →γ-Fe phase transformation that occurs in the steel [5]. As observed in Fig. 4a, the oxidation rate is more significant from 875 °C at temperatures high in the two phase field region. The reduction of the oxidation rate observed at 950 and 1050 °C is related to the detachment of the oxide from the steel. If transport of gas (CO_2, CO) is too slow or inhibited, then the pressure of gas would build up. Once it exceeds certain level, it would cause blistering or rupture of the scale decreasing the oxidation rate [5].

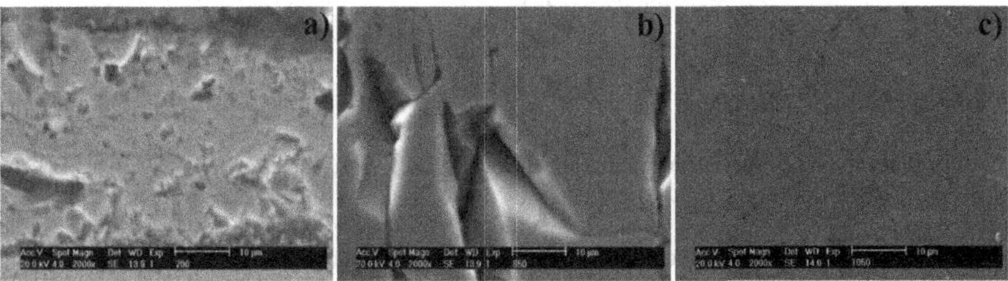

Figure 3.Microstructure of oxides formed during annealing of hot-rolled GNO electrical steel after 150 min at: a) 700 °C, b) 850 °C and c) 1050 °C.

Hence, the decrease or inhibition of decarburization can be attributed to both a transition from cracked to a crack-free structure and to the increase of the oxide film, which make carbon removal difficult. According to Fig. 1b, when annealing is performed at 875 °C and 900 °C, C content decreases with time only until 120 min and 90 min, respectively. Further annealing times does not have any effect on C concentration (curves E and F of Fig. 1b). Decarburization at higher temperatures does not occur even for short annealing times (curves G and H of Fig. 1b). These results suggest that the transition from a cracked to a crack-free oxide structure occurs faster at higher temperatures. This can explain the absence of decarburization at 950 and 1050 °C (curves G and H of Fig. 1b).

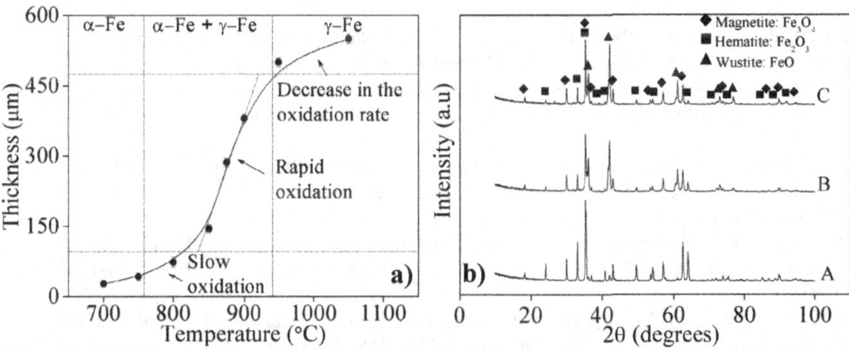

Figure 4. a) Thickness of the oxide layer as a function of temperature in samples annealed during 150 min and b) XRD patterns of oxides formed at: A) 700 °C, B) 850 °C and C) 1050 °C.

Fig. 4b shows the XRD patterns corresponding to oxides formed in the hot rolled electrical steel after 150 min of annealing at 700, 850 and 1050 °C. It is clearly observed that hematite is the main phase in the oxide formed at temperatures in the α-Fe phase field (700 °C). In this case, the amount of wustite present in the oxide layer is very small. When annealing temperature is increased to 850 and 1050 °C, the presence of hematite decreases and the amount of wustite increases. It is generally recognized that hematite and wustite behave as brittle and ductile phases during deformation at high temperature [5, 6]. Therefore, the present results suggest that formation of pores or cracks in the oxide are favored by the presence of a mixture of brittle/ductile phases in the oxide. In contrast, the formation of crack-free structures is related to the ductile behavior of the oxide due to the increase of wustite and the decrease of hematite in the oxide layer.

CONCLUSIONS

Decarburization of the experimental hot-rolled GNO electrical steel follows Wagner´s decarburization kinetics model only when thin and cracked-oxides structures are formed. Thick and crak-free oxide structures make difficult the C difussion resulting in slower decarburization rate than that predicted by the model. Pores and cracks are favored by the presence of a mixture of fragile and ductile phases in the oxide. The ductil behavior of wustite as well as the reduction of the most fragil phase (hematite) promotes the formation of crack-free oxides.

ACKNOWLEDGMENTS

A special acknowledgment is extended to CONACYT for the financial support under grant No. 166596. The contribution of Felipe Márquez, Socorro García, Sergio Rodríguez and Francisco Botello in this investigation is duly recognized.

REFERENCES

1. J. Gautam, Control of surface graded transformation textures in steels for magnetic flux carrying applications, PhD Thesis, Delft University of Technology, 2011.
2. L. Kestens, J. J. Jonas, P. Van Houtte and E. Aernoudt, Orientation selective recrystallization of nonoriented electrical steels, Metall. Mater. Trans., 27, 2347-2358 (1996).
3. E. J. Gutiérrez., A. Salinas, Effect of annealing prior to cold rolling on magnetic and mechanical properties of low carbon non-oriented electrical steels, J. Magn. Magn. Mater, 323, 2524-2530 (2011).
4. A. R Marder, S.M Perpetua, J.A. Kowalik and E.T. Stephenson, The effect of carbon content on the kinetics of decarburization in Fe-C alloys, Metall. Trans., 16A, 1160-1163 (1985).
5. R. Y Chen, W. Y. D. Yuen, Review of the high-temperature oxidation of iron and carbon steels in air or oxygen, Oxidation of metals, 59, 433-468 (2003).
6. S. Weihua, A study on the characteristics of oxide scale in hot rolling of steel, PhD thesis, University of Wollongong, 2005.

Mater. Res. Soc. Symp. Proc. Vol. 1485 © 2013 Materials Research Society
DOI: 10.1557/opl.2013.280

Effect of cooling rate on the formation and distribution of carbides in nodular iron alloyed with Cr

Ramses Zenil[1], Jose A. García[1], Gerardo A. Ruiz[1], Alejandro D. Basso[2], Jorge. Sicora[2]
[1]Departamento de Ingeniería Metalúrgica, Facultad de Química, Universidad Nacional Autónoma de México.
[2]INTEMA-Universidad Nacional de Mar del Plata, Argentina

ABSTRACT

This investigation deals with the effect of cooling rate on the formation of carbide nodular iron alloyed with 2.2 % Cr and equivalent carbon near 4.3%. In the experimental stage three Y-blocks of 1.5, 3 and 5.5 cm in thickness are poured in green sand molds. Castings are sectioned in two positions (wall and center samples) in order to determine the characteristics of the precipitated carbides (fraction, distribution and relative size) from surface to center and from the bottom to the top of the castings applying quantitative analysis of images. The obtained results show the presence of carbides in all of the castings. Finer carbides are obtained in the thinnest casting but with a high variation between the samples located in the wall and center. All castings present massive carbides in the last freezing zone (LFZ). Therefore the cooling rate associated with casting thick has an important effect on the fraction and distribution of carbides.

INTRODUCTION

Carbide nodular iron is a recent variation of nodular iron, nowadays it is studied to get Austenpered Carbide Ductile Iron (CADI) with high wear and abrasion resistance associated to presence of carbides in an ausferrite matrix. CADI´s materials have important applications in mining, construction, agricultural or in components with a high demand for abrasion and wear use. The Carbide Ductile Iron (CDI) is produced by casting processes that require controlling the chemical composition and cooling rate during solidification [1, 2]. Nodular iron produced with high cooling rate promotes a high undercooling and the material phase decomposition follows the Fe-carbide system and the precipitation of graphite is reduced. Some alloying elements decrease the difference between the eutectic temperature of the Fe-C diagram under stable and metastable conditions [3] promoting the precipitation of carbides and graphite at the same time. The presence of whitening elements such as Cr, V, Mo and B affect the microstructure, they are strong carbide promoters, however they can also cause a heterogeneous distribution of carbides and strong microsegregation patterns in the Last Freezing Zones (LFZ) [4]. Thermal stability of as cast carbides is important during the austenitizing stage of the CADI´s isothermal heat treatment. Thus the objective of this work is to determine the effect of three cooling rates on the average local carbide fraction, carbide mapping (distribution), overall carbides % in each sample and the comparative size in nodular irons cast in green sand moulds.

EXPERIMENT

The nodular irons are manufactured in an induction furnace with a 120 Kg of capacity, raw materials are steel scrap with high manganese and scrap of cast iron. The chemical composition adjustments are made with FeSi75, FeMn70, pure Cr, FeMo74 and pure copper, graphite is used as recarburant agent. The inoculation is performed with FeSi75 and nodulizing treatment with FeSi45Mg6. Y block castings of 1.5, 3 and 5.5 cm thick are poured in green sand molds. Two samples of each casting located one of wall and another in the center are cut out, see Figure 1a. Each sample is analyzed in three positions: wall, intermediate and center (horizontal direction) and also three position in the vertical direction: bottom, center and top, see Figure 1b. At each position three images are taken to obtain the average local carbides %, with the purpose to get the distribution of carbides in zones with high (wall sample) and low cooling rate (center sample) in the same casting. Samples are etched with a fresh solution of 10 ml HNO_3, 4 ml HF and 87 ml H_2O. They are immersed during 1 to 2 minutes at room temperature. The fraction of carbides is determined by image processing using Image Pro Plus software at 100x. Distribution mappings of carbides through out the two samples for each alloy are obtained. The cooling rate of each casting is obtained by thermal analysis.

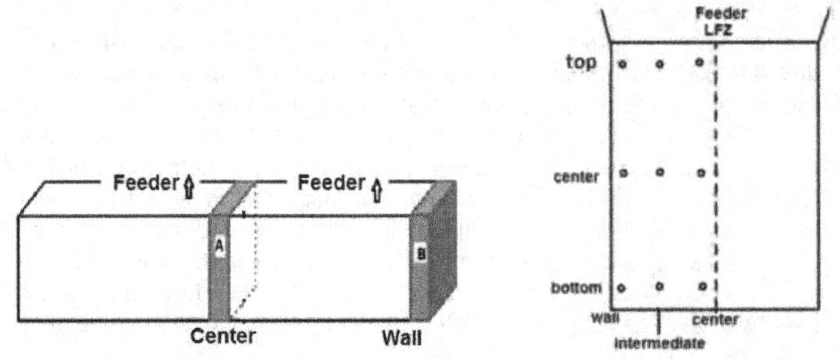

Figure 1. (a) Samples cut from the Y block casting: center and wall. (b) Punctual zones analysis in each sample from the bottom to top (vertical analysis) and wall to center (horizontal analysis). LZF is the Last Freezing Zone.

DISCUSSION

The characteristics of the graphite are shown in Table I, the matrix in all cases is perlite and carbides. The results of the carbide mapping and the carbide local average fraction (%) in the horizontal (wall to center) and vertical (bottom to top) directions for both of the samples wall and center are shown in figures 2a and 2b for a 1.5 cm thickness, figures 3a and 3b for a 3 cm thickness and figures 4a and 5b for a 5.5 cm thickness. Table II shows the difference of the average fraction of carbides between wall and center samples for each casting as well as the difference between the overall fraction carbide between wall and center samples.

Table I. Characteristics of graphite nodules as a function of thickness of the casting

	Nodule characteristic [5]		
Casting thickness (cm)	Nodularity % (AFS)	Nodule Size (AFS)	Nodule Count (AFS)
1.5	90	7-8	200
3.0	90	7	150
5.5	90	6-7	100

Table II. Average carbide fraction wall-center samples, difference between the overall fraction carbide (%) as a function of the sample position (wall and center).

Thikcness (cm)	Overall average carbide fraction of wall-center samples %	Average fraction carbide (%)			
		Wall sample	Center sample	Difference	% Difference
1.5	24.00	28.50	19.50	9.00	32.78
3.0	16.6	17.70	15.50	2.20	12.42
5.5	22.35	22.80	21.90	0.90	3.90

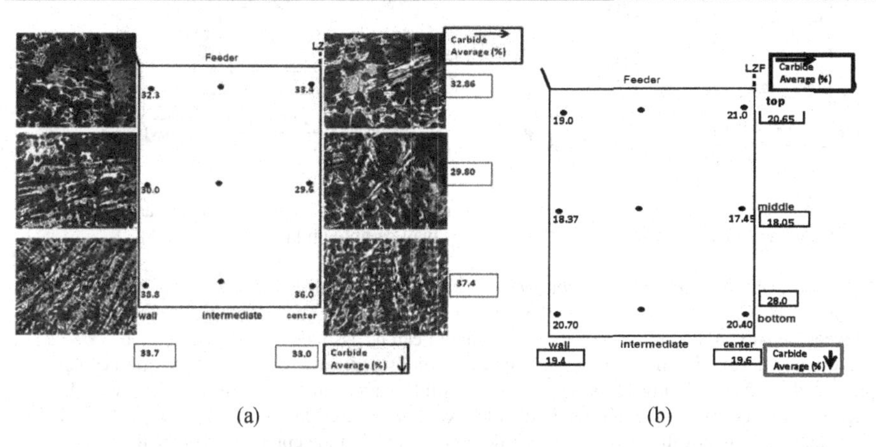

(a) (b)

Figure 2. Average local fraction carbides and average carbide % in the vertical and horizontal directions in the casting with 1.5 cm thickness (a) wall sample showing the carbide mapping microstructures and (b) center sample just with average local carbide %.

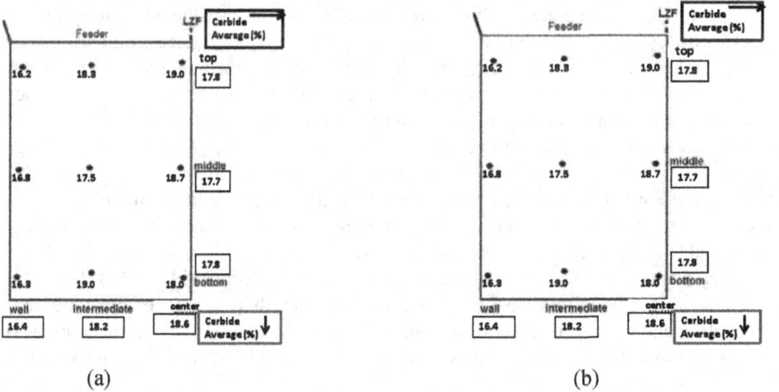

(a) (b)

Figure 3. Average local fraction carbides and average carbide % in the vertical and horizontal directions in the casting with 3.0 cm thickness (a) wall sample and (b) center sample.

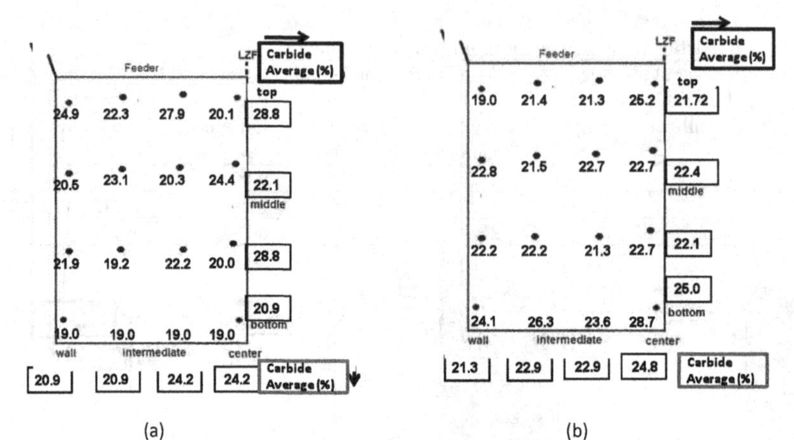

Figure 4. Average local fraction carbides and average carbide % in the vertical and horizontal directions in the casting with 5.5 cm thickness (a) wall sample and (b) center sample.

The obtained cooling rates for each casting from the cooling curve are 1.87 °C/sec for the 1.5 cm, 0.88 °C/sec for the 3 cm and 0.35 °C/sec for the 5.5 cm thick casting. The differences between density and nodule count shown in Table I can be associated to the nucleation conditions imposed by the cooling rate. For a low cooling rate, there are few nucleation centers promoted by the small undercooling, the higher solidification time allows growth of few and large graphite nodules, as observed, in the matrix of the produced cast iron. Figures 2, 3 and 4 show a heterogeneous distribution of carbides between wall and center samples for the three castings. This carbide distribution variation is associated to the local cooling conditions and also indicates a chromium segregation phenomenon during the solidification. The Casting with the smallest thickness (1.5 cm) has the highest local carbide fraction and the overall carbide fraction (Table II). This behavior is associated with the presence of Cr that is a strong whitening element in cast iron and reduces the difference of the eutectic temperatures between the iron-carbon stable and metastable diagrams [2]. In other words chrome promotes the undercooling required for carbide precipitation and the effect is enhanced by the high cooling rate associated with the casting thickness. For this same casting there is a significant difference between the fraction of carbides precipitated at the wall and in the center. This indicates that the effect of local cooling rate is not eliminated by the addition of 2.2% Cr. Also Table II shows that the 3.0 and 5.5 cm thick castings have a lower fraction of carbides. This suggests that the addition of chrome promotes carbide precipitation in thicker castings, but there is a competition between reducing stable and metastable eutectic temperatures and the slow cooling rate imposed by the thickness of the casting. This results in a reduction of the amount of precipitated carbides. Thus the presence of 2.2% Cr does not eliminate the effect of cooling rate, there is a dependency on the local rate of cooling and carbide fraction from zone to zone of the casting (wall-center samples). Furthermore an irregular distribution of carbides in the vertical and horizontal directions is also observed.

The average total carbide variations in wall-center segments shown in Table III are 2.2 for the 3 cm thick casting, 0.9 for the 5.5 cm thick and 9 for the 1.5 cm thick casting. Such results indicate that thicker castings have the most homogeneous distribution of carbides. Overall fractions of carbides present in the three castings agree in magnitude with the values reported by Laino et al [5] for the case of nodular iron directionally solidified with contents of 2 and 2.5% Cr. Three K-blocks show ledeburitic carbides, although there are areas in which carbides are observed continuously. The size of the carbides is finer and better distribution in the rapid cooling zones of the casting, that is in areas close to the wall, while the Last Freezing Zones (LFZ) are massive and heavy. This effect is associated with the local solidification rate and conditions of iron carbide nucleation when it solidifies in the metastable Fe-C diagram. The results show carbide variations between neighboring areas. These features are intrinsic to solidification of nodular iron under heterogeneous condition, making it difficult to establish the conditions for completely homogeneous distribution of carbides in nodular iron whitening with this level of chromium.

CONCLUSIONS

Nodular irons with carbides are obtained associated with the addition of 2.2% Cr and cooling rate conditions from 28 to 17% of carbides. The thinnest casting (1.5 cm) shows the highest average carbide fractions at 28.5 % in areas located at the wall to 19.5 % in areas located in the center of the casting. The overall average carbide fraction is 24 %. Chrome is a strong carbide forming element but it does not eliminate the effect of the local cooling rate. Carbide variations are less significant in heavy section of the casting, they are 0.9 for the 3 cm thick casting, 2.2 for the 5.5 cm thick casting. The carbide distribution is less heterogeneous in this case, but with minor carbide fractions. Carbides size in the thinnest casting (1.5 cm) is comparatively smaller than for the heavy casting where massive carbides are precipitated in the last freezing zone of all the castings (LFZ). The fraction of precipitated carbides depends of the competition between the cooling rate and whitening potential of the alloying element.

ACKNOWLEDGMENTS

We thank PAPIIT IT118411-3 for the financial support granted for this study, as well as IQM E, Cándido Atlatenco T. fir his support in the manufacture of nodular iron and IQM Victor A. Aranda Villada for his help in the metallographic evaluation of carbides.

REFERENCIAS.

1. F. Binsczki, A. Kowalsky, J. Furmanec, Vol. 7, Issue 2, 2007, pp. 115-118.
2. G.M Goodrich, Cast Iron Microstructures Anomalies and their causes, AFS Transaction 1997-30, pp. 669.
3. Z. Jiyang, Dalian University of Technology China, Serial Report February 2009
4. M. Lagarde, A. Basso, R.C. Dommarco, J. Sikora, ISIJ International Vol. 51 (2011), No. 4, pp. 645-650.
4. American Foundrymen´s Society (AFS), "Foundrymen´s guide to ductile iron microstructures" Des Plaines, Illinois, (1984).
5. S. Laino, R. Dommarco, J.A. Sikora, Desarrollo de fundiciones nodulares austemperadas con carburos CADI, Congreso SAM-CONAMET , Mar del Plata Arg. Octubre 2005.

Mater. Res. Soc. Symp. Proc. Vol. 1485 © 2013 Materials Research Society
DOI: 10.1557/opl.2013.281

Effects of Austenitizing Temperature and Cooling Rate on the Phase Transformation Texture in Hot Rolled Steels

N.M. López [1], and A. Salinas R[2]
Centro de Investigación y de Estudios Avanzados del Instituto Politécnico Nacional, Saltillo Campus, P.O. Box 663, Saltillo Coahuila, México 25900.
e-mails: [1]nan_mar0904@hotmail.com, [2]armando.salinas@cinvestav.edu.mx

Keywords: Texture, phase transformation, steels, EBSD, orientation relationships

ABSTRACT

The effects of austenitizing temperature and cooling rate on the microstructures and textures produced by phase transformations in high strength hot rolled Fe-C-Mn steel plates are investigated using orientation imagining microscopy. Samples machined from the plates are austenitized at temperatures between 820-950°C during 30 minutes and quenched in either iced-water, water or oil. Finally, the quenched samples are tempered at 450°C during 30 minutes. Characterization of microstructure and textures produced by these heat treatments was performed by conventional metallography using a reflected light microscope and orientation imaging microscopy using backscattered-electron diffraction patterns in a scanning electron microscope with thermo-ionic electron source.

The results show that the microstructure and texture produced under a given combination of austenitizing temperature and cooling rate are strongly dependent on the mechanism involved in the phase transformation of the austenite (γ). High austenitizing temperatures and cooling rates promote martensitic transformation and development of textures containing significant volume fractions of Br, Cu, transformed-Cu and transformed-Br orientation components. In this case, the austenite and martensite phases are clearly related through the Kurdjumov-Sachs orientation relationship. In contrast, low temperatures and low cooling rates result in a complex mixture of transformation products, such as polygonal ferrite, Widmanstäten ferrite, martensite, bainite and pearlite. The textures formed under these conditions are quite different and contain significant volume fractions of cube, rotated-cube, Goss and rotated-Goss components, following the Bain orientation relationship.

INTRODUCTION

During processing of steel, deformation and thermal gradients affect the microstructure, texture and mechanical properties of the final product. Since the development of texture is due to a complex interaction between at least three processes, deformation, recrystallization and phase transformation, a great deal of empirical knowledge is need to understand, control and generate, under processing conditions, the necessary microstructures for specific applications [1].
Phase transformation in metals and alloys involves a change in crystalline structure, e.g. bcc to fcc, which involves the movement and displacement of atoms. In steels, the phase transformation from austenite to ferrite, pearlite, bainite or martensite is of great technological importance. A typical feature of phase transformations is the appearance of orientation relationships [2]. During hot rolling of high strength steels a deformation texture is developed in the parent phase (γ) which is later inherited by the transformation products depending on conditions and operating parameters. Bain, Kurdjumov-Sachs and Nishiyama-Wasaerman have proposed models that

describe phase transformations in steels using geometric relationships between the original crystals and transformed crystals [3-6].
When hot-rolled steel is quenched after an austenitizing heat treatment, the martensitic phase transformation takes place ($\gamma \rightarrow \alpha'$). This transformation occurs following the Kurdjumov-Sachs orientation relationship. In contrast, when the cooling rate after the austenitizing treatment is slow, the transformation microstructure is a complex mixture of micro-constituents and the resulting texture can be explained in terms of the Bain orientation relationship [7, 8].

In the present work the effects of austenitizing temperature and cooling rate on the microstructures and textures produced by phase transformation in high strength hot rolled Fe-C-Mn steels are investigated. The experimental results show that when the steel is cooled rapidly from $T>Ac_3$ causes the phase transformation from $\gamma \rightarrow \alpha'$, generating textures with transformation components such as Cu, Br, transformed-Cu and transformed-Br, following the Kurdjumov-Sachs orientation relationship. In contrast, quenching from $T<Ac_3$ and slow cooling rates result in mixtures of transformation products (polygonal ferrite, Widmanstäten ferrite, martensite, bainite and perlite) with textures containing significant volume fractions of cube, Goss, rotated-cube and rotated-Goss components following the Bain orientation relationship.

EXPERIMENTATION

Experimental samples in the present work were obtained from 7 mm thick, high strength hot rolled Fe-C-Mn steel plate. The chemical composition of the plate was determined by optical emission spectrometry (wt%C=0.25, wt%Mn=1.20, wt%Si=0.23, wt%Cu=0.01, wt%Ti=0.024, wt%Al=0.13 y wt%B=0.006, wt%Fe=balance). Samples, 15 mm long and 7 mm wide, were cut to perform thermal treatments. The samples were austenitized at temperatures between 820-950°C during 30 minutes and quenched using three different quenching media (iced-water, water and oil). Finally, the quenched samples were tempered at 450°C during 30 minutes. Characterization of microstructure and textures produced by these heat treatments were performed by conventional metallography using a reflected light microscope and orientation imaging microscopy using backscattered-electron diffraction patterns obtained in a scanning electron microscope with thermo-ionic electron source (OIM-EBSD/MEB).

RESULTS AND DISCUSSION

Microstructure and texture of hot rolled steel

The microstructure of the hot rolled steel is show in Fig. 1a, which is composed of equiaxed ferrite grains approximately 15 μm in size, and perlite islands arranged in bands aligned parallel to the hot rolling direction. This type of phase distribution affects the homogeneity of austenite during austenitization, the different micro-constituents have different transformation temperatures and therefore affect the phase transformation [9]. Fig. 1b shows the corresponding orientation map of the ferrite phase. As can be appreciated, the texture of the ferrite is mainly random but, as shown in the φ_2=45° section of the ODF in Fig. 1c, the orientation distribution exhibits five weak orientation components (Cu, transformed-Cu, Br, Goss and rotated-Goss). The presence of these texture components suggests that the phase transformation after hot rolling took place from partially recrystallized austenite. It is noteworthy that the relatively larger orientation density of the Br and Cu components can be attributed to the relatively large Mn content and the presence of B in the present steel. Kim et al. [10] showed that Mn and B have a strong influence on the formation of Br and Cu components as a result of the transformation of austenite. As the content of these elements increases, the orientation density of the Br and Cu components increase while the orientation density of the rotated-cube components usually formed as a result of the transformation of recrystallized austenite decrease.

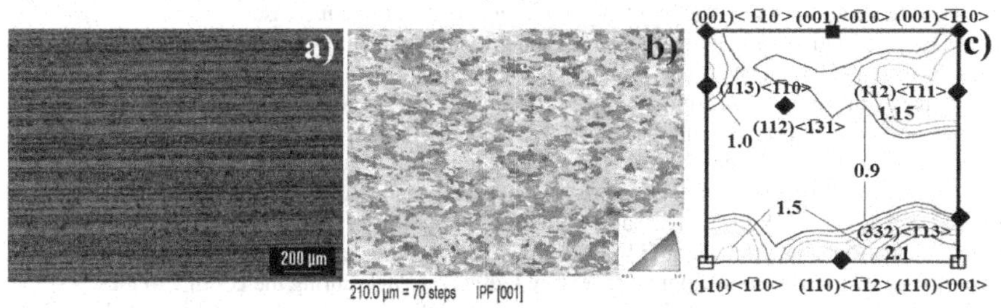

Fig. 1. a) Microstructure, b) orientation map and c) φ_2=45° section ODF of the hot rolling steel

Effects of temperature and cooling rate on the microstructure and phase transformation texture

The effects of quenching temperature and cooling rate on the microstructures of the investigated steel are presented in Fig. 2. As can be seen, in the case of samples quenched from 820°C the microstructures consists of a mixture of transformation products: polygonal ferrite, Widmanstäten ferrite, perlite, bainite and martensite depending on the cooling rate. It is evident that quenching the steel from T<Ac_3 causes that the transformation ocurs from a mixture of austenite and ferrite.

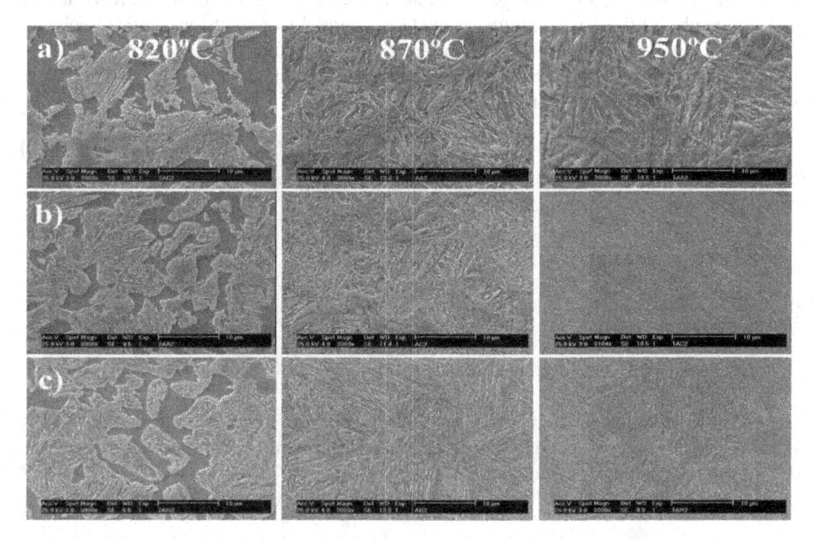

Fig. 2. Photomicrographs of samples quenched from 820° C, 870° C and 950 °C and cooled in a) oil, b) water, c) iced-water, and tempered at 450°C during 30 minutes

In contrast, when quenching is performed from 870°C and 950°C the microstructure consists of martensite with a lat-type morphology. The main difference in the microstructures produced by different quenching tempertures and cooling rates is the martensite packet size, which becomes finer as the quenching temperature and cooling rate increases.
In phase transformations involving nucleation and growth, the crystal orientations produced are always defined at the nucleation stage, although some variation in the growth rate could affect the presence of the orientation components. In most cases, the nucleation is dominated by processes occurring at grain boundaries. Ferrite nuclei are oriented such that they have a semi-coherent low energy interface with at least one of the adjacent austenite grains. In the case of steels, the observed orientation relationships are not unique and occupy a range of values between Bain, K-S y N-W [3-6]. The same conditions apply to martensite even though in this case the nucleation is intra-granular [11]. When the steel is treated at high temperatures, austenite (parent phase) acquires a crystallographic texture that is inherited during the cooling to the transformation products (ferrite, bainite or martensite). ODF´s corresponding to the treated samples at 820°C and quenched in different media are shown in Fig. 3. As can be seen, quenching from this temperature produces relatively strong textures with maximum orientation densities of about 10 mrd. Four main orientation components are observed, Goss, rotated-Goss, cube and rotated-cube. The presence of the first two components in materials with phase transformation indicates that austenite was recrystallized prior to the transformation. The Bain orientation relationships is defined as a 45° rotation about a <110> axis common in each crystal (parent and doughter) with the following correspondence relation:

$\{001\}^{\gamma}//\{001\}^{\alpha}$ $<100>^{\gamma}//<110>^{\alpha}$

The main texture components produced in the ferrite by transformation of recrystallized austenite with a cube {001}<010> texture are the Goss, rotated-Goss, cube and rotated-cube components (see Fig. 3d). Additionally other orientation components can be present in the transformed ferrite (e.g. Cu, Br) since to the cube component in the austenite can also generate them as a result of the transformation [9, 7].

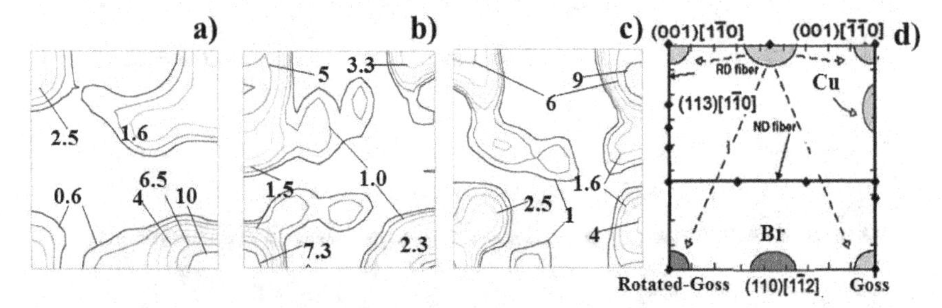

Fig. 3. φ_2=45° ODF section steel quenched from 820°C in a) oil, b) water, c) iced-water and tempered at 450°C. d) φ_2=45° section ODF of Euler space showing bcc texture components formed from the fcc cube component [8].

Martensite grows by a displacive mechanism which results in shear deformation in addition to the volume change that accompanies the transformation and the presence of residual stresses [11]. The growth of this phase is known to follow very closely the K-S orientation relationship given by [4]:

$(111)^{\gamma}//(110)^{\alpha}$ $[\bar{1}10]^{\gamma}//[\bar{1}11]^{\alpha}$, or as: $(111)^{\gamma}//(110)^{\alpha}$ $[1\bar{1}2]^{\gamma}//[\bar{1}12]^{\alpha}$

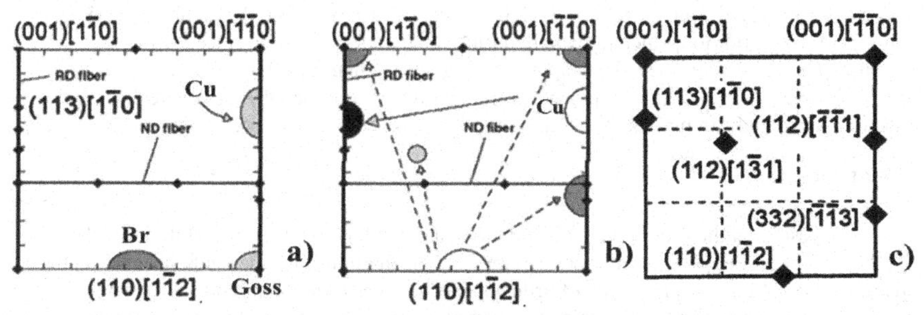

Fig. 4. a) Deformed austenite components, b) orientation components resulting of the phase transformation austenite to martensite, and c) φ_2=45° section ODF showing the typical martensitic orientation components [8, 12].

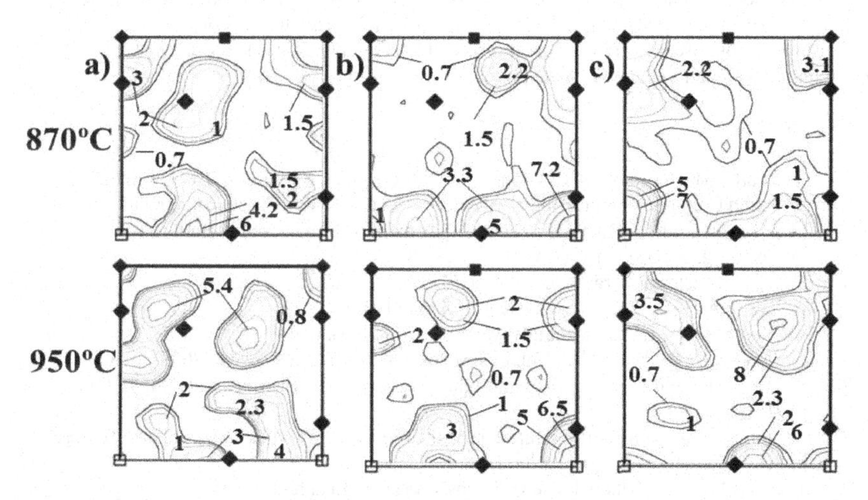

Fig. 5. φ_2=45° section ODF of treated steel at 870°C and 950°C during 30 minutes, quenched in different medias a) oil, b) water y c) iced-water and tempered at 450°C during 30 minutes.

Fig. 4a shows the typical ideal orientations present in deformed austenite and Fig. 4b illustrates the ideal orientations produced by the martensitic phase transformation. In contrast to the phase transformation from recrystallized austenite, this type of transformation is more complex, since the texture of deformed austenite consists of several parent orientations who are responsible of a number of orientation component in the product phase. Figs. 4c and 5 show the martensite texture components formed following the K-S orientation relationship and the ODF´s of samples quenched from 870°C and 950°C, respectively. In contrast to the textures observed in samples quenched from $T<Ac_3$ (Fig. 3) in these ODF´s predominate the Cu, transformed-Cu, Br, and transformed-Br components. The presence of these components indicate that phase transformation occurred from austenite which developed a strong Cu and Br components (Fig. 4a) during soaking prior to quenching which, during the transformation produced strong transformed-Cu and transformed-Br components (Fig. 4b). When the steel is quenched from 950°C these texture components are present with high intensity in the ODF´s. For steels applications were high strength is required, such orientation components are of great importance.

CONCLUSIONS

The results of the present investigation indicate that the texture of austenite formed during heating of (α-Fe+pearlite) microstructures in a B-containing steel depends strongly on the temperature. At $T<Ac_3$, the presence of ferrite appears to favor formation of a cube-type texture in the austenite that transforms on cooling to a complex microstructure consisting of ferrite, bainite and martensite with a relatively strong texture characterized by Goss, rotated-Goss, cube and rotated-cube components.

In contrast, on heating to $T>Ac_3$ the austenite develops Cu and Br orientations which, on cooling, transform to transformed-Cu, transformed-Br following the K-S orientation relationship. In this case, the effect of cooling rate affects mainly the packet size of the lath-type morphology developed in the martensite phase.

REFERENCES

1. A. Roatta, M. Bertinetti, A.L. Foourty, C. Sobrero, J.W. Signorelli, R.E. Bolmaro. Brazilian Congress Engineering and Materials Science. Vol.17 (2006), p. 6444.
2. M.P Butrón Guillén. Doctoral Thesis. Department of Mining and Metallurgical Engineering, McGill University. (1995), p. 18.
3. E.C. Bain. Materials Transactions. Vol.70 (1924), p. 25.
4. G. Kurdjumov, G. Sachs. Z. Physical. Vol. 64 (1930), p. 325.
5. G. Wassermann. Arch. Eisenhüttenwes. Vol.16 (1933), p. 347.
6. L. Sandoval, H. Urbassek, P. Entel. New Journal of Physics. Vol.11 (2009), p. 30.
7. S. Morito, H. Saito, T. Ogawa, T. Furuhara, T. Maki. ISIJ International. Vol.45 (2005), p. 91.
8. J.J. Jonas. Proceedings of the International Conference on Microstructure and Texture in Steels and other Materials. (2008), p. 3.
9. T. F. Majka, D. K. Matlock and G. Krauss. Metallurgical and Materials Transaction.Vol.33 (2002), p. 1627.
10. S.I. Kim, S.H. Choi, Y.C. Yoo. Materials Science Forum. Vol.495 (2005), p. 537.
11. B. Hutchinson, L. Ryde, P. Bate. Materials Science Forum. Vol. 495 (2005), p.1141.
12. C.S DA Costa Viana, M.P. Butrón Guillén, J.J. Jonas. Textures and Microstructures. Vol. 26 (1996), p. 599.

Mater. Res. Soc. Symp. Proc. Vol. 1485 © 2013 Materials Research Society
DOI: 10.1557/opl.2013.282

Fracture behavior of heat treated liquid crystalline polymers

A Reyes-Mayer [1], B Alvarado-Tenorio [1], A Romo-Uribe[1*], O Flores[1], B Campillo[1] and M Jaffe [2]
[1] Laboratorio de Nanopolimeros y Coloides, Instituto de Ciencias Fisicas, Universidad Nacional Autonoma de Mexico, Cuernavaca, Mor. 62210, MEXICO.
[2] New Jersey Institute of Technology, Newark NJ, U.S.A.
* To whom correspondence should be addressed: aromo-uribe@fis.unam.mx

ABSTRACT

Thermotropic polymers are thermally treated in air at temperatures T_a, where $\Delta T = T_a - T_{s\rightarrow n}$=40°C, and $T_{s\rightarrow n}$ is the solid-to-nematic transition. Samples are extruded thin films of a series of thermotropic random copolyesters termed B-N, COTBP and RD1000. The thermal treatment produces a second endotherm without changing $T_{s\rightarrow n}$ for B-N and RD1000. However, for COTBP $T_{s\rightarrow n}$ is significantly increased. Regardless of the complex thermal behavior exhibited by the thermotropes, the thermal treatment produces a significant increase in Young's modulus, more than 30% for B-N and over 100% for COTBP. The increase in mechanical modulus is correlated with a thermally-induced fiber-like morphology.

INTRODUCTION

Thermotropic polymers exhibit high mechanical properties and excellent chemical and thermal resistance, making them suitable for high performance applications in automotive, electronic and aerospace industries [1-2]. Many thermotropic liquid crystalline polymers (LCPs) have been synthesized, but their mechanical properties are often poor at elevated temperatures. For example, copolymers of HBA and HNA show a 75% drop in tensile modulus on heating from room temperature to 150°C [3].

Several groups have investigated dynamic mechanical and dielectric properties of LCPs. The influence of monomer concentration, type of monomers, and the effect of temperature treatment on the mechanical relaxations of thermotropic LCPs have also been studied [4-6]. Moreover, the influence of degree of molecular orientation on mechanical properties has also been investigated [5].

The tensile properties of thermotropic LCPs can be improved by heat treatment. The optimal annealing temperature has been reported to be between 10 and 30°C below the melting point [7]. In a previous report on a series of thermotropic random copolyesters we have shown that thermal treatment indeed produces an increase in tensile modulus [8]. In this work we further investigate the influence of thermal treatment on microstructure using small-angle light scattering (SALS), X-ray scattering and scanning electron microscopy (SEM).

EXPERIMENTAL PROCEDURE

Materials. Extruded thin films of three thermotropic liquid crystalline copolyesters have been kindly supplied by the former Hoechst Celanese Research Corp. (Summit, NJ, USA). These are termed B-N, COTBP and RD1000. B-N is a copolyester based on random units of ~73 mol % B and ~27 mol % N. The chemical formula is shown in Figure 1a. This material is marketed by Ticona under the tradename of *Vectra A*®. COTBP (Figure 1b) is a random copolyester made of 66 mol% 1,4-hydroxybenzoic acid (B), 5 mol%2,6-hydroxynaphthoic acid (N), and 17.5 mol% (equal parts) of terephthalic acid (TA) and biphenol (BP). Finally, RD1000 is a wholly aromatic thermotropic LCP whose chemical composition is very similar to COTBP, however its specific chemical composition is proprietary information of Celanese Co.

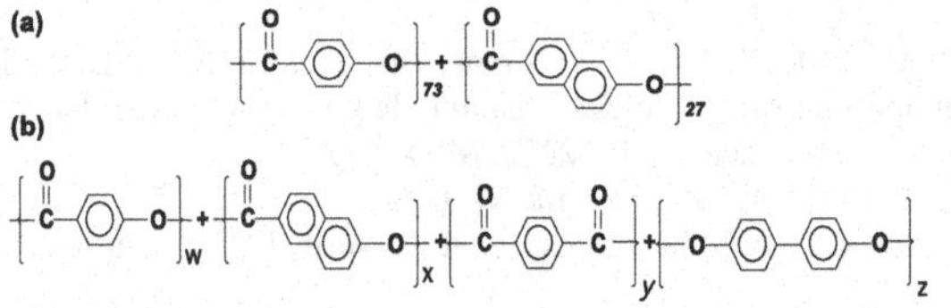

Figure 1. Chemical formulas of thermotropic LCPs. (a) B-N, (b) COTBP.

Thermal analysis. The thermal transitions are determined by differential scanning calorimetry (DSC). For DSC experiments, a DSC6000™ calorimeter manufactured by Perkin Elmer (Connecticut, USA) is used. Temperature and enthalpy calibration are carried out using analytical grade indium (T_m = 156.6 °C) and corrections are made for the instrument 'baseline'. The thermal transitions are determined at a heating rate of 20 °C/min.

Small-angle light scattering, SALS. Small-angle light-scattering patterns in H_V polarization condition are obtained using an in-house instrument equipped with a He-Ne laser (λ=632.8 nm) and described in detail elsewhere [9]. The light source is a vertically polarized He-Ne laser (wavelength λ=632.8 nm) of 0.8 mW power (model 1500 manufactured by JDS Uniphase Corp., Santa Rosa California, USA). The 0°-90° orientation of the polarizer and analyzer sets the so-called H_V polarization condition. For image recording a charge-coupled device (CCD, model PC-23C, Super Circuits, Taiwan) with a resolution of 200 μm/pixel is used.

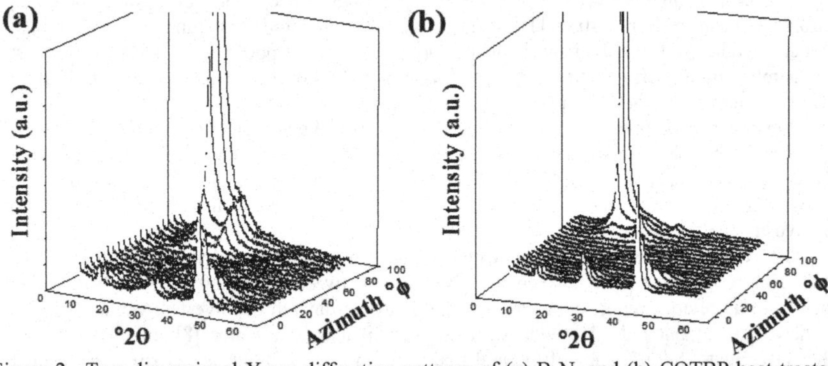

Figure 2. Two-dimensional X-ray diffraction patterns of (a) B-N, and (b) COTBP heat treated fibers. Graphite monochromator and CuKα radiation. Azimuthal angle ϕ=0° corresponds to meridional scan.

Mechanical properties. Mechanical properties are studied under uniaxial tension. Tests are carried out at room temperature on a universal testing machine Instron 4206 (USA) with a crosshead speed of 5 mm/min.

Wide-angle X-ray scattering. Two-dimensional wide-angle X-ray scattering (WAXS) patterns are obtained using a S-Max3000 rotating anode generator (Rigaku Inc.). This equipment employs Cu Kα (λ=1.5405 Å) as the radiation source, and is operated at 45 kV and 0.88 mA. The patterns are recorded using a flat-plate camera and Fuji image plates, with a sample-to-detector distance of 6 cm. The patterns are analyzed using the software POLAR v2.6 (Stonybrook Technology Inc., Stonybrook, NY).

SEM. The evaluation of the fracture mode in the tapes is carried out utilizing scanning electron microscopy (SEM). A SEM model JEOL JSM-5900LV is used. The fracture surfaces are metal coated with Au/Pd by a sputtering procedure.

RESULTS AND DISCUSSION

The as-extruded B-N thermotropic copolyester fibers exhibit aperiodic meridional reflections, consistent with reports by Blackwell *et al* and Windle *et al* [2, 10]. Figure 2a shows the corresponding wide-angle X-ray scattering pattern. COTBP (4 monomers copolyester) exhibits crystalline reflections and aperiodic meridional reflections, suggesting the existence of non-periodic layer crystalline structure, as in B-N [2] (see Fig. 2b).

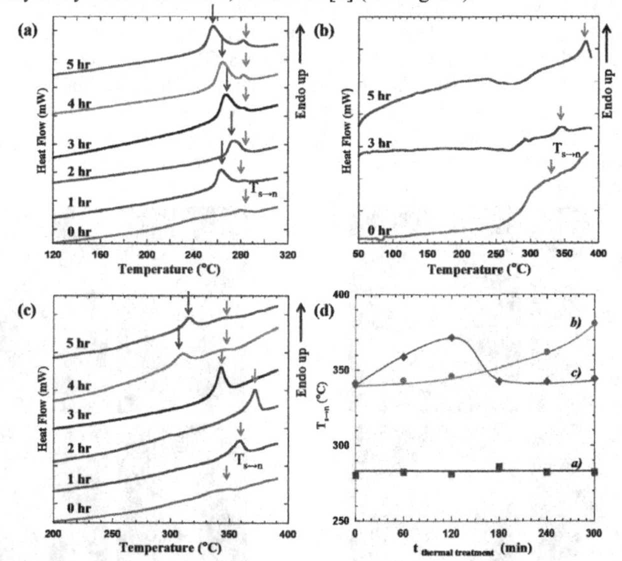

Figure 3. DSC heating traces as a function of heat treatment time for (a) B-N annealed at 240 °C, (b) COTBP annealed at 300 °C, and (c) RD1000 annealed at 300 °C. (d) $T_{s\rightarrow n}$, as a function of thermal treatment time for (a) B-N, (b) COTBP, and (c) RD1000.

DSC shows a solid-to-nematic transition temperature, $T_{s\rightarrow n}$ of 280°C, 340°C and 341°C for B-N, COTBP and RD1000, respectively for as-extruded films. Thus, heat treatment is applied, for a period of several hours under air conditions, at 240°C for B-N, and 300°C for COTBP and RD1000, that is, at $\Delta T=T_a-T_{s\to n}\sim 40$°C. Figure 3a-c shows DSC heating traces for the thermotropes, and Figure 3d shows a plot of $T_{s\to n}$ as a function of heat treatment time. Note that for B-N, $T_{s\to n}$ is independent of the thermal treatment. However, for COTBP and RD1000, $T_{s\to n}$ increases, at least initially for RD1000. The thermal treatment also produces a second endotherm for B-N and RD1000. Previously, Chung *et al* and Romo-Uribe *et al* had reported the production of a second endotherm in B-N[11-13].

The extruded films of COTBP and RD1000 are optically transparent, suggesting a well aligned texture, i.e., a monodomain. Moreover, it is found that thermal treatment influences the optical texture in COTBP and RD1000. Figures 4 and 5 show SALS patterns (H_V polarized) of COTBP and RD1000 as a function of heat treatment time. The extrusion direction is along the horizontal thus the diamond shaped patterns show that the defect texture mostly (disclinations) is aligned along the extrusion axis. The azimuthal spread of intensity in COTBP suggests that the texture is less well aligned than for RD1000. Moreover, heat treatment increases the alignment of the defect texture in RD1000 as evidenced from the reduction of azimuthal spread of intensity.

Uniaxial tensile testing is carried out, stretching at 5 mm/min at room temperature. Results show a significant increase in mechanical modulus (E) after the films are heat treated. However there is a brittle behavior of the LCP films with a steep linear elastic region [8]. Figure 6 shows a series of SEM micrographs. In the as received-condition the samples exhibit different fracture modes. B-N shows again a brittle type fracture (Fig. 6a), COTBP has a ductile fracture (Fig. 6c), and RD1000 shows again a brittle fracture (Fig. 6e). This behavior can be associated to E values previously obtained of 72 GPa, 60 GPa and 108 GPa respectively [8]. The ductile behavior for COTBP can be associated to the lower E value. The higher E values for B-N and RD1000 apparently produced a brittle fracture mode. After 4 hours treatment, the E values increase up to 90 GPa (Fig. 6b), 152 GPa (Fig. 6d) and 140 GPa (Fig. 6f) respectively. Consequently the fracture mode becomes mainly brittle for all treated samples. Additionally, this type of fracture shows a characteristic fiber-like morphology.

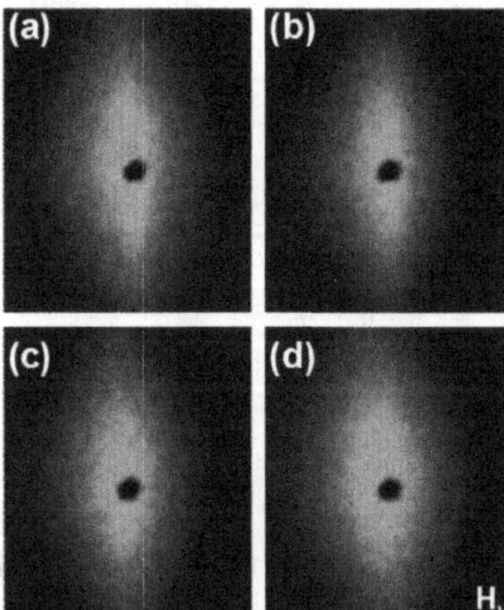

Figure 4. SALS patterns of COTBP heat treated at 300°C for (a) 0, (b) 60, (c) 240, and (d) 300 min, in air. Extrusion axis is horizontal.

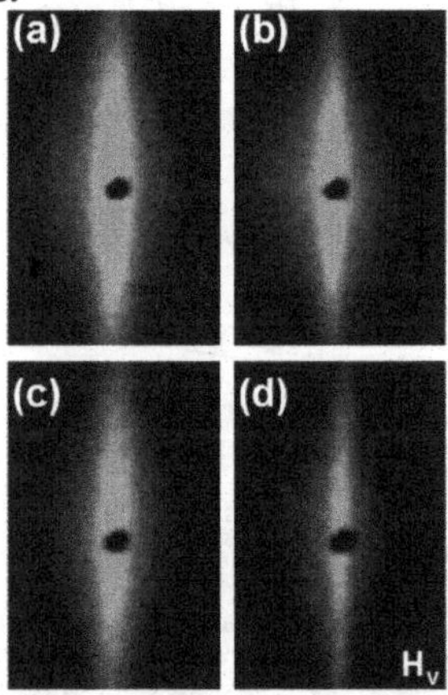

Figure 5. SALS patterns of RD1000 heat treated at 300°C for (a) 0, (b) 60 (c) 120, and (d) 180 min, in air. Extrusion axis is horizontal.

The influence of thermal treatment is also investigated via wide-angle X-ray scattering. Figure 7 shows patterns for films as-extruded and heat-treated for up to 4 hours. The as-extruded samples are highly oriented along the extrusion axis, especially COTBP and RD1000, thus correlating the low defect density (Figures 4 and 5) with high molecular orientation (Figures 7 c and e). After heat treatment, the equatorial (horizontal axis) and meridional (vertical axis) crystalline reflections become very sharp and well-defined, especially the off-equatorial reflections, suggesting the elimination of defects in the crystal structure. The sharpening of crystalline reflections is more evident for COTBP (Figures 7 c, d), and these results correlate with the increase in mechanical modulus after heat treatment.

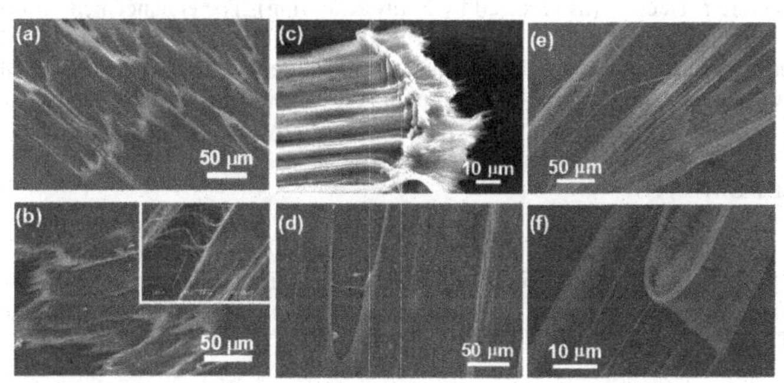

Figure 6. SEM micrographs of as-received (upper row) and heat-treated 4 hrs (lower row) fractured films. (a, b) B-N, annealed at 240 °C, (c, d) COTBP, and (e, f) RD1000, annealed at 300 °C.

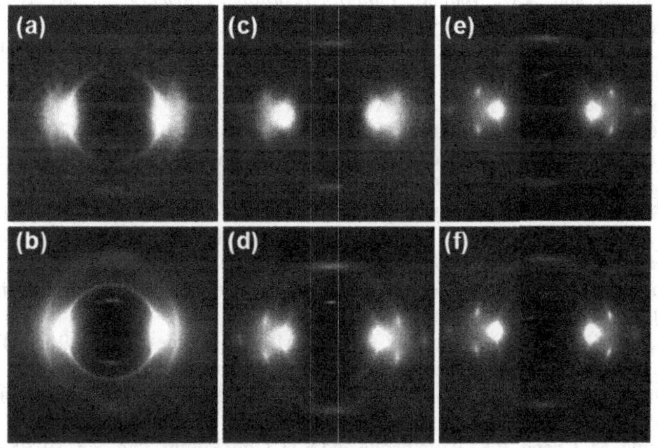

Figure 7. WAXS patterns of as-received (upper row) and heat-treated 4 hrs (lower row) extruded films. (a, b) B-N, annealed at 240 °C, (c, d) COTBP, annealed at 300 °C, and (e, f) RD1000, annealed at 300 °C. CuKα radiation. Extrusion axis is vertical.

CONCLUSIONS

The results of this investigation show that thermal treatment below the solid-to-nematic transition temperature produces significant changes in microstructure, and thermal and mechanical properties of thermotropic LCPs. The transition temperatures of thermotropic LCPs are significantly influenced by thermal treatment. There is an increase of solid-to-nematic transition ($T_{s \to n}$) or the production of a second endotherm, when thermal treatment is carried out at $\Delta T = T_a - T_m \sim 40°C$, T_a being the treatment temperature. The results also show that thermal treatment induced the elimination of optical defects (as revealed by SALS), and an increase in crystalline structure perfection (as revealed by X-ray scattering). The enhancement of crystalline structure and elimination of texture defects produce, therefore, an increase of the Young's modulus. Thermal treatment also produces a change in fracture mode, from ductile to brittle. The brittle fracture mode displays a fiber-like morphology, as revealed by SEM, and this fracture behavior is correlated with the crystalline perfection induced by the thermal treatment.

ACKNOWLEDGMENTS

AR-M gratefully acknowledges the financial support of DGAPA-UNAM. Thanks to Hoechst Celanese Corp (Summit, NJ, USA) for providing extruded films. Thanks to Mr. I. Puente for technical support (FQ-UNAM). This research is financed by CONACyT, CIAM2008, grant 107294, and DGAPA-UNAM, PAPIIT, grant IN109810.

REFERENCES

1. L.C. Sawyer, H.C. Linstid and M. Romer, Plastics Engineering 54, 37 (1998).
2. A.M. Donald and A.H. Windle, *Liquid Crystalline Polymers*, 2nd ed. (Cambridge: Cambridge University Press, 1992).
3. M. Cakmak, A. Teitge, H.G. Zachmann and J.L. White, J. Polym. Sci. Poly. Phys. 31, 371 (1993).
4. A. Romo-Uribe, Proc. R. Soc. Lond. A457, 207 (2001).
5. T.L.D. Collins, G.R. Davies and I.M. Ward, Polym. Adv. Tech. 12, 544 (2001).
6. A. Romo-Uribe, B. Alvarado-Tenorio, M.E. Romero-Guzman, L. Rejon and R. Saldivar-Guerrero, Polym. Adv. Techn. 20, 759 (2009).
7. S. Salahshoor-Kordestani, S. Hanna and A.H. Windle. Polymer. 41, 6619 (2000).
8. A. Reyes-Mayer, A. Constant, A. Romo-Uribe and M. Jaffe. Mater. Res. Soc. Symp. Proc. 1373 DOI: 10.1557/opl.2012.317 (2012).
9. A. Romo-Uribe, B. Alvarado-Tenorio and M.E. Romero-Guzmán, Rev. LatinAm. Metal. Mat. 30, 190 (2010).
10. J. Blackwell, G.A. Gutierrez and R.A. Chivers, Macromolecules. 17, 1219 (1984).
11. A. Romo-Uribe, T.J. Lemmon and A.H. Windle, J. Rheol. 41, 1117 (1997).
12. T.S. Chung, M. Cheng, S.H. Goh, M. Jaffe and G.W. Calundann, J. Appl. Polym. Sci. 72, 1139 (1999).
13. T.S. Chung, M. Cheng, P.K. Pallathadka and S.H. Goh, Polym. Eng. Sci. 39, 953 (1999).

Mater. Res. Soc. Symp. Proc. Vol. 1485 © 2013 Materials Research Society
DOI: 10.1557/opl.2013.283

Dynamically recrystallized austenitic grain in a low carbon advanced ultra-high strength steel (A-UHSS) microalloyed with boron under hot deformation conditions

I. Mejía[1], E. García-Mora[1], G. Altamirano[1], A. Bedolla-Jacuinde[1] and J. M. Cabrera[2,3]
[1] Instituto de Investigaciones Metalúrgicas, Universidad Michoacana de San Nicolás de Hidalgo. Edificio "U", Ciudad Universitaria, Morelia, Michoacán, México.
[2] Departament de Ciència dels Materials i Enginyeria Metal·lúrgica, ETSEIB – Universitat Politècnica de Catalunya. Av. Diagonal 647, Barcelona, Spain.
[3] Fundació CTM Centre Tecnològic, Av. de las Bases de Manresa, 1, Manresa, Spain.

ABSTRACT

This research work studies the dynamically recrystallized austenitic grain size (D_{rec}) in a new family of low carbon NiCrCuV advanced ultra-high strength steel (A-UHSS) microalloyed with boron under hot deformation conditions. For this purpose, uniaxial hot-compression tests are carried out in a low carbon A-UHSS microalloyed with different amounts of boron (14, 33, 82, 126 and 214 ppm) over a wide range of temperatures (950, 1000, 1050 and 1100°C) and constant true strain rates (10^{-3}, 10^{-2} and 10^{-1} s^{-1}). Deformed samples are prepared and chemically etched with a saturated aqueous picric acid solution at 80°C in order to reveal the D_{rec} and examined by light optical (LOM) and scanning electron microscopy (SEM). The D_{rec} is related to the Zener-Hollomon parameter (Z), and thereafter the D_{rec} divided by Burger's vector (b) is related to the steady state stress (σ_{ss}) divided by the shear modulus (μ) (Derby model). Results shown that the D_{rec} in the current steels is fine ($\approx$ 23 μm) and almost equiaxed, and the recrystallized grain size-flow stress relationship observed after of plastic deformation is consistent with the general formulation proposed by Derby. It is corroborated that boron additions to the current A-UHSS do not have meaningful influence on the D_{rec}.

INTRODUCTION

Recent years have seen many developments in steel technology and manufacturing processes to build vehicles of reduced weight and increased safety. For this purpose, Advanced High Strength Steels (AHSS) have been developed. The AHSS include newer types of steels such as dual phase (DP), transformation-induced plasticity (TRIP), complex phase (CP), B steels (BS), and martensitic steels (MART), which are primarily multi-phase steels, and contain ferrite, martensite, bainite, and/or retained austenite in quantities sufficient to produce outstanding mechanical properties [1,2]. Researchers have studied the B effect in steels for a long time [3-7], particularly because of its potential to increase steel hardenability. Nowadays, B-bearing steels are extensively applied in the construction of heavy machinery, building structures, marine platforms and pipelines [8-11]. It is well-known that B atoms segregate towards austenite grain boundaries and increase hardenability of steel by suppressing the nucleation of ferrite [12]. The grain size is the only microstructural parameter that can simultaneously increase the strength and toughness levels of steels [13]. Dynamic recrystallization (DRX) is the process of formation of new small and dislocation free grains from deformed material. On the other hand, deformation conditions, temperature and strain rate, have strong effects on the D_{rec} [14,15]. Derby and Ashby [16-18] have shown a "universal" relation between σ_{ss} and D_{rec}. It is worth noting that the D_{rec} is independent of the initial grain size, and exclusively dependent on the σ_{ss}. At present, there are just a few studies strictly focused on the B effect on the D_{rec} of advanced high strength steels. This research work studies the boron effect on the D_{rec} of a new family of low carbon NiCrVCu advanced ultra-high strength steels (A-UHSS).

EXPERIMENTAL DETAILS

The present experimental low carbon NiCrVCu advanced ultra-high strength steels (A-UHSS) are melted in the Foundry Laboratory of the Metallurgical Research Institute-UMSNH using high purity raw materials in a 25 kg capacity induction furnace. Table I shows the chemical composition of the six experimental steels examined in this study.

Table I. Chemical composition of low carbon NiCrVCu advanced ultra-high strength steels (A-UHSS) microalloyed with B (wt. %).

A-UHSS	C	Mn	Si	S	Cu	Cr	Ni	V	Al	N	B
B0	0.15	0.40	0.42	0.02	0.52	1.31	2.44	0.22	0.0026	0.0091	**0**
B1	0.12	0.40	0.40	0.02	0.51	1.31	2.38	0.22	0.0040	0.0100	**0.0014**
B2	0.11	0.41	0.43	0.01	0.46	1.33	2.26	0.24	0.0048	0.0082	**0.0033**
B3	0.11	0.40	0.35	0.02	0.51	1.30	2.37	0.22	0.0030	0.0086	**0.0082**
B4	0.10	0.40	0.33	0.01	0.49	1.30	2.30	0.22	0.0036	0.0079	**0.0126**
B5	0.10	0.41	0.32	0.02	0.50	1.30	2.42	0.22	0.0031	0.0087	**0.0214**

Isothermal uniaxial hot-compression tests are carried out over a wide range of temperatures (950, 1000, 1050 and 1100°C) and constant true strain rates (10^{-3}, 10^{-2} and 10^{-1} s^{-1}). Specimens are quenched in water immediately after the compression tests. Deformed samples are prepared and chemically etched with a saturated aqueous picric acid solution at 80°C in order to reveal the D_{rec}, and examined by light optical (LOM) and scanning electron microscopy (SEM). Average D_{rec} is measured by image analysis using digitalized pictures through Sigma Scan Pro software. First, the D_{rec} is related to the Z, and thereafter the D_{rec} divided by Burger's vector (b) is related to the σ_{ss} divided by shear modulus (μ) (Derby model).

RESULTS AND DISCUSSION

Hot flow curves

Figure 1 shows the hot flow curves of the low carbon advanced ultra-high strength steel (A-UHSS) microalloyed with boron for B0 and B5 steels as a function of the strain rate at 950 and 1100°C. These flow curves exhibit the expected behavior when DRX occurs [14,15]. All steels show a similar hot flow behavior, with the classic dependence of peak stress (σ_p) and peak strain (ε_p) on temperature and strain rate, i.e., σ_p and ε_p increase as strain rate increases and temperature decreases. In all cases, the steel microalloyed with boron (B5) shows lower σ_p values than the steel without boron (B0). Similar remarks can be done about the σ_{ss}. It is also observed that the softening beyond the σ_p value is more pronounced in the B5 steel.

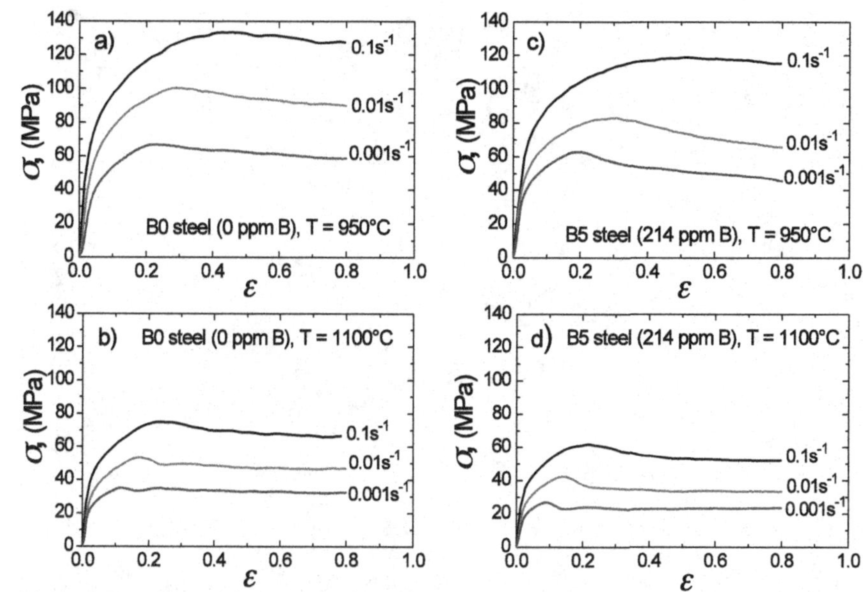

Figure 1. Experimental flow curves obtained by uniaxial compression tests for B0 and B5 steels as a function of the strain rate at 950 and 1100°C.

Dynamically recrystallized austenitic grain (D_{rec}) and Derby model

Figure 2 shows examples of D_{rec} for B0, B1, B3 and B5 steels. It can be seen that the austenitic grains are small and almost equiaxed. DRX is essentially a process of limited growth of the new recrystallized grains, with repeated nucleation. Figure 3 shows SEM micrographs of austenite grains of B1 and B3 steels at 1050°C and 0.1 s^{-1}, which both prior austenitic grain boundaries and transformed microstructure can be seen in detail. This microstructure is mainly composed of bainite and martensite (B+M) with lath structure morphology. The micrographs clearly identify the product as the martensitic phase formed at the grain boundary and triple points. Figure 4 shows the D_{rec} as a function of the Z. It can be seen that the D_{rec} ranging between 12 and 45 μm, with average size of 23 μm. It is worth mentioning that the D_{rec} tends to remain constant as Z is increased (i.e., as the strain rate is increased and/or the temperature lowered), although a slight increase can be noted with the highest Z values. On the basis of these results, it is evident that boron additions to the current steels do not have significant influence on the D_{rec}. Figure 5 shows the relation between the D_{rec} and the effective σ_{ss} for B0 and B5 steels, and the constants of Derby´s equation are summarized in table II. In general the exponent of the D_{rec} (m) is close to 2/3, and the constant n value ranging between 5.6×10^{-11} and 3.11×10^{-8}, which is consistent with the general formulation proposed by Derby. It is worth noting that the D_{rec} is independent of initial grain size and it remains constant during the steady state regime [19].

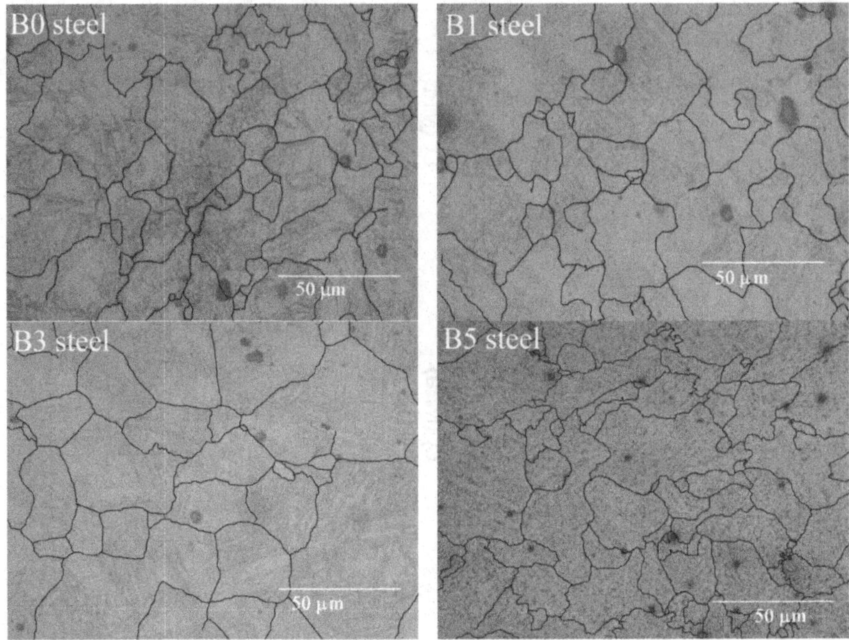

Figure 2. Optical micrographs of dynamically recrystallized austenitic grain (D_{rec}) of B0, B1, B3 and B5 steels at 1100°C and 0.001 s^{-1}.

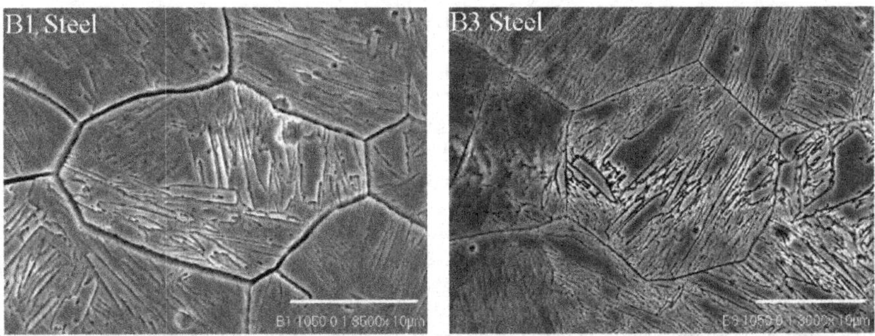

Figure 3. SEM micrographs of dynamically recrystallized austenitic grain (D_{rec}) of B1 and B3 steels at 1050°C and 0.1 s^{-1}, showing packets and blocks within individual prior austenite grains.

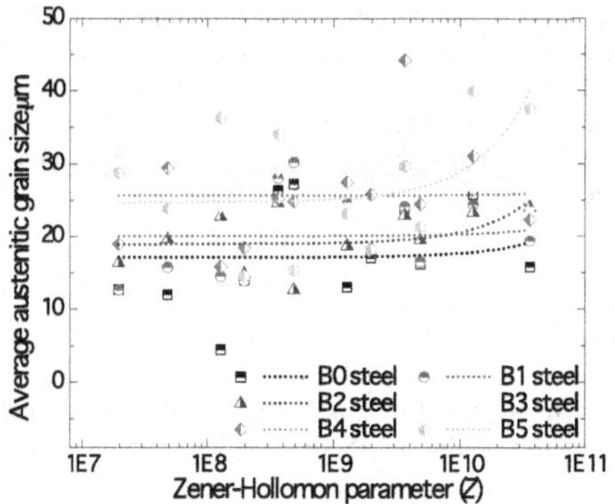

Figure 4. Dynamically recrystallized austenitic grain size versus Zener-Hollomon parameter (Z).

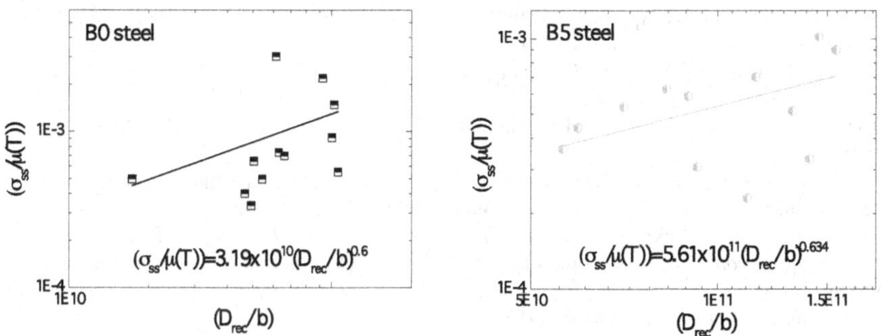

Figure 5. Dynamically recrystallized grain size (D_{rec}) divided by Burger's vector (b) as a function of the steady state stress (σ_{ss}) divided by shear modulus (μ).

Table II. Constants of Derby's equation.

A-UHSS Steel	Derby´s equation: $\frac{\sigma_{ss}}{\mu(T)}=n\left(\frac{D_{rec}}{b}\right)^{-m}$	
	n	m
B0	$3.19x10^{-10}$	0.600
B1	$3.11x10^{-8}$	0.397
B2	$6.78x10^{-11}$	0.640
B3	$1.15x10^{-10}$	-0.297
B4	$1.50x10^{-10}$	0.506
B5	$5.61x10^{-11}$	0.634

CONCLUSIONS

The present D_{rec} in the current low carbon NiCrCuV advanced ultra-high strength steels (A-UHSS) microalloyed with boron under uniaxial hot-compression conditions is fine and almost equiaxed (ranging between 12 and 45 μm and with average size of 23 μm). D_{rec} tends to remain constant as Z increases (i.e., strain rate increases and/or temperature decreases), although a slight increase can be noted with the highest Z values. In general, the recrystallized grain size-flow stress relationship observed after of plastic deformation is consistent with the general formulation proposed by Derby. It is corroborated that boron additions to the current steels do not have meaningful influence on the D_{rec}.

ACKNOWLEDGMENTS

E. García-Mora would like to thank CONACYT (México) for the scholarship support during this project. Authors also acknowledge CMEM-UPC (Spain), for the support and technical assistance in this research work. Funding is obtained through project CICYT-MAT2008-06793-C02-01 (Spain) and CIC-UMSNH (México).

REFERENCES

1. Committee on Automotive Applications, International Iron & Steel Institute, *Advanced High Strength Steel Application Guidelines* 1-9 (2006).
2. H. T. Jiang, D. Tang and Z. L. Mi, *J. Iron Steel Res.* **19**, 1-6 (2007).
3. K. A. Taylor and S. S. Hansen, *Metall. Mater. Trans.* **A21**, 1697-1708 (1991).
4. H. Tameiro, M. Murata, R. Habu and M. Nagumo, *Trans. Iron Steel Inst. Jpn.* **27**, 120-129 (1987).
5. J. E. Morral and T. B. Cameron, in *Boron in Steel* edited by S. K. Banerji and J. E. Morral, (The Metallurgical Society of AIME, Milwaukee, USA, 1980) pp. 19-32.
6. R. Habu, M. Miyata, S. Sekino and S. Goda, *Trans. Iron Steel Inst. Jpn.* **18**, 492-500 (1978).
7. D. H. Werner, *Boron and Boron Containing Steels*, 2nd ed. (Verlag Stahl Eisen, Dusseldorf, 1995) pp. 15-20.
8. B. M. Kapadia, *J. Heat Treat.* **5**, 41-53 (1987).
9. C. J. Heckmann, D. Ormston, F. Grimpe, H. G. Hillenbrand and J. P. Jansen, *Ironmaking Steelmaking* **32**, 337-341 (2005).
10. H. J. Jun, J. S. Kang, D. H. Seo, K. B. Kang and C. G. Park, *Mater. Sci. Eng.* **A422**, 157-162 (2006).
11. H. Kagechika, *ISIJ Int.* **47**, 773-794 (2007).
12. X. M. Wang and X. L. He, *ISIJ Int.* **42**, 38-46 (2002).
13. Y. Ohmori and K. Yamanaka, in *Boron in Steels* edited by S. K. Banerji and J. E. Morral, (Metall. Soc. AIME, New York,1980) pp. 44-60.
14. T. Sakai and J.J. Jonas, *Acta Metall.* **32**, 189 (1984).
15. F.J. Humphreys and M. Hatherly, *Recrystallization and Related Annealing Phenomena,* (Pergamon Press, Oxford, 1995).
16. B. Derby and M. F. Ashby, *Scripta Metall.* **21**, 879-884 (1987).
17. B. Derby, *Acta Metall.* **39**, 955-962 (1991).
18. B. Derby, *Scripta Metall. Mater.* **27**, 1581-1586 (1992).
19. H. J. McQueen, S. Yue, N. D. Ryan and E. Fry, *J. Mater. Process. Technol.* **53**, 293-310 (1995).

Mater. Res. Soc. Symp. Proc. Vol. 1485 © 2013 Materials Research Society
DOI: 10.1557/opl.2013.284

Magnetization study of the kinetic arrest of martensitic transformation in as-quenched $Ni_{52.2}Mn_{34.3}In_{13.5}$ melt spun ribbons

F.M. Lino-Zapata[1], J.L. Sánchez Llamazares[1], D. Ríos-Jara[1], A.G. Lara-Rodríguez[2], and T. García-Fernández[3]

[1] Instituto Potosino de Investigación Científica y Tecnológica, Camino a la Presa San José 2055 Col. Lomas 4ª, San Luis Potosí, S.L.P. 78216, México.
[2] Instituto de Investigaciones en Materiales, UNAM, Circuito Exterior s/n, Ciudad Universitaria, México D.F. 04510, México.
[3]Universidad Autónoma de la Ciudad de México, Prolongación San Isidro 151,Col. San Lorenzo Tezonco, México DF, C.P. 09790, México

ABSTRACT

The kinetic arrest of martensitic transformation (MT) has been observed in as-solidified $Ni_{52.2}Mn_{34.3}In_{13.5}$ melt spun ribbons. The main characteristics of this unusual field-induced magneto-structural phenomenon have been determined through a dc magnetization study. The sample studied was fabricated by rapid solidification using the melt spinning technique at a high quenching rate of 48 ms^{-1}. At room temperature, it is a single phase austenite (AST) with the bcc B2-type crystal structure and Curie temperature of T_C^A=285 K. With decreasing temperature, the austenite phase transforms into the martensite phase (MST) with $T_C^M \approx$185 K at a starting martensitic transition temperature of M_S=275 K. A moderate but progressive kinetic arrest of the AST to MST transformation has been observed for magnetic field values above H=10 kOe and was studied up to H_{max}= 90 kOe. The metastable character of the non-equilibrium field-cooled state is revealed by the decreasing behavior of the saturation magnetization under a large magnetic field of 50 kOe after temperature cycling from 10 K to 150 K. The total magnetization difference $\Delta\sigma$ between the zero field-cooling and field-cooling pathways of the temperature dependence of magnetization shows irreversible and reversible components and the former decreases with decreasing temperature.

INTRODUCTION

In the last few years, considerable attention has been paid to the investigation of the physical phenomena related to the first-order structural martensitic transition in ferromagnetic shape memory alloys of the ternary alloy systems Ni-Mn-X (X= Sn, In, Sb). The most important of these are magnetic superelasticity [1], the giant inverse magnetocaloric effect [1-4], and large magneto-resistance [5-7].

In connection with the structural transition, the most important phenomenon induced by the magnetic field in these materials is the field-induced reverse martensitic transformation [8, 9]. However, in some bulk Ni-Mn-In based alloys [10-13] and melt spun Mn-Ni-In alloy ribbons [14] the kinetic arrest of the martensitic transformation has been reported (i.e., a volume fraction of austenite remains frozen into an equilibrium martensitic matrix when the material is cooled below the martensitic final structural transition temperature M_f under the application of a static magnetic field beyond a certain critical value). In this paper, we report the occurrence of this unusual phenomenon in as-quenched melt spun ribbons of the ferromagnetic shape memory Heusler alloy, $Ni_{52.2}Mn_{34.3}In_{13.5}$. Several dc magnetization studies have been carried out in order to reveal the distinct features of this uncommon effect.

EXPERIMENTAL SECTION

A bulk alloy of nominal composition $Mn_{50}Ni_{36}In_{14}$ was prepared by arc melting in argon atmosphere. The sample was remelted several times to ensure good homogeneity, while Mn evaporation losses were carefully compensated by adding the corresponding excess of this element to maintain the 50:36:14 composition. From this as-cast alloy, melt spun ribbons were obtained in a highly pure argon environment at a tangential speed of the copper wheel of 48 ms^{-1} using a home-made single-roller melt spinner apparatus.

The ribbon samples so fabricated were studied by means of X-ray diffraction (XRD), differential scanning calorimetry (DSC), scanning electron microscopy (SEM), and magnetization measurements. DSC measurements were performed in a TA INSTRUMENTS model Q200 under a high purity Ar flow at a heating and cooling rate of 3 °C/min. XRD diffractograms were recorded on powdered samples in a BRUKER D8 ADVANCE powder diffractometer (Cu-Kα; $20^\circ \leq 2\theta \leq 90^\circ$; step increment: 0.02°). SEM studies were carried out in FEI XL 30 SEM with an EDAX microanalysis system. The average elemental chemical composition was estimated by energy dispersive spectroscopy (EDS). Magnetization measurements were carried out with a vibrating sample magnetometry (VSM) in a Quantum Design PPMS® EverCool®-9T platform. The magnetization as a function of temperature curves $\sigma(T)$ were measured in zero field-cooling (ZFC), field-cooling (FC), and field-heating (FH) modes at a heating/cooling rate of 1.0 K/min.

RESULTS AND DISCUSSION

Figure 1(a) shows the room temperature XRD pattern measured on powdered as-solidified ribbons The diffraction peaks were satisfactorily indexed on the basis of a bcc B2-type crystal structure with a lattice parameter *a* of 0.2993 nm. The crystallization of AST in this ordered crystal structure, instead of in the highly ordered $L2_1$-type structure, has been frequently reported in as-solidified Ni-Mn-In alloy ribbons with similar composition [15, 16]. The inset in Figure 1 shows the typical cross sectional microstructure of these ribbons. Their average thickness is ~9-10 um, while the microstructure consists of well-formed columnar microcrystalline grains that grew along the whole thickness of the ribbons. The average elemental chemical composition determined by EDS after more than 30 analyses is $Ni_{52.2}Mn_{34.3}In_{13.5}$ (0.1 at. %) which differs from the nominal one which is $Ni_{50}Mn_{34}In_{16}$ (analyses were performed on both surfaces of ribbon pieces and on their cross section). This difference in composition is mainly attributed to the losses in Mn due to evaporation and reaction of the alloy with the quartz crucible during the radio-frequency induction melting process used to fabricate the melt spun ribbons.

Figure 1(b) shows the FH and FC $\sigma(T)$ curves measured under a low static magnetic field of 50 Oe (i.e., $\sigma(T)^{50Oe}$), together with the DSC scans. The extrapolation procedure employed to estimate the characteristic temperatures of the direct and reverse martensitic transformation are illustrated in the $\sigma(T)^{50Oe}$ curves. The values obtained by both methods, together with the Curie temperature of AST and MST (referred to as T_C^A and T_C^M, respectively) are listed in Table I.

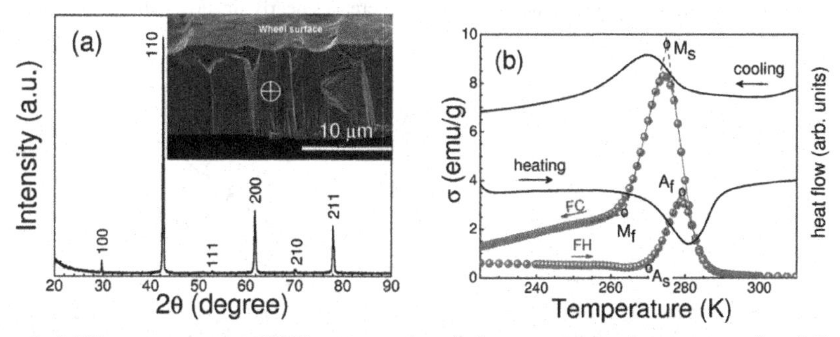

Figure 1. (a) Room temperature XRD pattern and typical cross section microstructure (inset) for the sample studied. The crossed circle in the inset indicates the rolling direction. **(b)** $\sigma(T)^{50Oe}$ curves measured in field-heating (FH) and field-cooling (FC) regimes (red dots) and DSC scans (black line) for the sample studied. The extrapolation procedure used to determine the phase transition temperatures is illustrated in the $\sigma(T)^{50Oe}$ curve.

Table I. Starting and finishing temperatures of the forward and reverse first-order martensitic transformation, thermal hysteresis (determined as $\Delta T = A_f - M_S$), Curie temperature of AST (T_C^A) and MST (T_C^M) determined from the $\sigma(T)^{50Oe}$ and DSC curves.

Method used	A_S [K]	A_f [K]	M_S [K]	M_f [K]	ΔT [K]	T_C^A [K]	T_C^M [K]
DSC	267	290	281	255	9	-	-
$\sigma(T)^{50Oe}$	271	279	275	264	4	285	257

The temperature dependence of the magnetization at 50 kOe $\sigma(T)^{50kOe}$ is shown in Figure 2(a). The large jump of magnetization observed when the structural transformation from MST to AST, and vice versa, takes place is due to the higher saturation magnetization of AST with respect to MST. In comparison with other ferromagnetic shape memory alloys in the Ni-Mn-In system, the atypical behavior observed in the graph is the magnetization difference between the ZFC and FC (and FH) pathways of the curve in the martensitic existence region (hereinafter referred to as $\Delta\sigma^{FC\text{-}ZFC}$) which is better shown in the inset. This phenomenon is known as the kinetic arrest of the martensitic transformation. It has been reported in some specific bulk Ni-Mn-In based alloys, such as $Ni_{50}Mn_{36}In_{14}$ [10, 11], $Ni_{45}Co_5Mn_{36.7}In_{13.3}$ [12], and $Ni_{50}(Mn,Fe)_{34}In_{16}$ [13], and melt spun $Mn_{50}Ni_{40}In_{10}$ alloy ribbons [14]. When the temperature decreases, a volume fraction of AST does not transform into MST but remains frozen within the equilibrium martensitic matrix as a result of the applied magnetic field. Hence, the kinetic arrest of MT gives rise to a magnetically inhomogeneous system in which magnetic MST and AST phases coexist. The resulting FC state should be metastable, and therefore if H is removed or thermal fluctuations are introduced, the system should tend to the ZFC equilibrium state. In the present alloy, the kinetic arrest of MT has been detected for magnetic field values above 10 kOe. Figure 2(b) shows the field dependence of $\Delta\sigma^{FC-ZFC}$ up to a maximum magnetic field of 90 kOe at different temperatures in the martensitic existence region. As can be seen, the fraction of AST arrested, which is roughly proportional to $\Delta\sigma^{FC-ZFC}$, increases nearly linearly with the magnetic field strength.

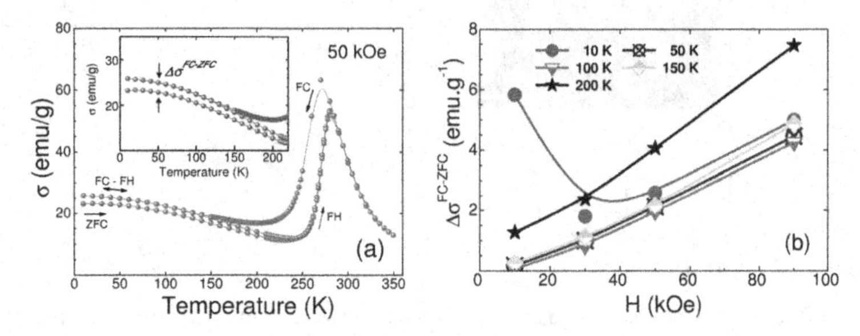

Figure 2. (a) $\sigma(T)$ curves measured at 50 kOe in ZFC, FC, and FH modes. Inset: zoom into the $\sigma(T)^{50\ kOe}$ curve in the martensitic region that shows the magnetization difference between the ZFC and FC pathways of the curve. **(b)** Field dependence of the difference $\Delta\sigma^{FC-ZFC}$ at different temperature in the martensitic region.

To prove the metastable nature of the FC inhomogeneous state, two different magnetization experiments were carried out. In the first one, thermal energy fluctuations were introduced in the martensitic region through successive thermal cycling starting from the field-cooled state, i.e., the sample is field-cooled under a large magnetic field of 50 kOe to 10 K, and the temperature increased from 10 K to successively increasing maximum temperatures T_{max} of 50 K, 100 K, and 150 K (the measured $\sigma(T)$ curves in Figure 2 are referred to as s1, s2, and s3; after the sample reaches the respective T_{max} value the temperature is decreased to 10 K). The measured $\sigma(T)$ curves are shown in Figure 3(a); the inset zooms into the low temperature region to show better the decreasing behavior of $\sigma(T)$. The metastable character of the FC state is manifested by the magnetization decrement after successive temperature cycling since the thermal fluctuations gradually transform a small fraction of AST into MST, indicating that the system tends to the equilibrium ZFC state [10, 14]. In a second experiment, the sample is again cooled-down under the same field value until two selected temperatures are reached, namely 10K and 200 K, and the field is removed to check if the FC magnetization state tends to the ZFC one. Figure 3(b) compares the thermo-remanent demagnetization isotherms $\sigma(H)^{TR-1}$ which are followed by field-up and field-down isotherms (numbered as 2 and 3, or $\sigma(H)^{TR-2}$ and $\sigma(H)^{TR-3,}$ respectively; blue circles), with the magnetization isotherms measured from the thermally demagnetized state in increasing and decreasing the field (denoted as $\sigma(H)^{ZFC}$; red stars). As shown, there is a difference between $\sigma(H)^{TR-1}$ and $\sigma(H)^{ZFC}$ at H_{max}= 50 kOe and T= 10 K, as well as at T= 200 K (which is better displayed in the right insets of the graphs). The coincidence of $\sigma(H)^{TR-1}$ and $\sigma(H)^{ZFC}$ at 200 K indicates that when the field is removed the frozen fraction of AST fully transforms into MST as has been previously reported for bulk $Ni_{50}Mn_{36}In_{14}$ [10, 11] and melt spun $Mn_{50}Ni_{40}In_{10}$ alloys [14]. However, the nearly complete overlap of $\sigma(H)^{TR-1}$ and $\sigma(H)^{TR-2}$ at 10 K implies that after the field removal AST remains practically untransformed. Hence, in the presently studied alloy, the magnetization change shows reversible and irreversible components. The latter, i.e., $\Delta\sigma(H)^{irrev}$, is indicated in the right insets of Figure 3(b).

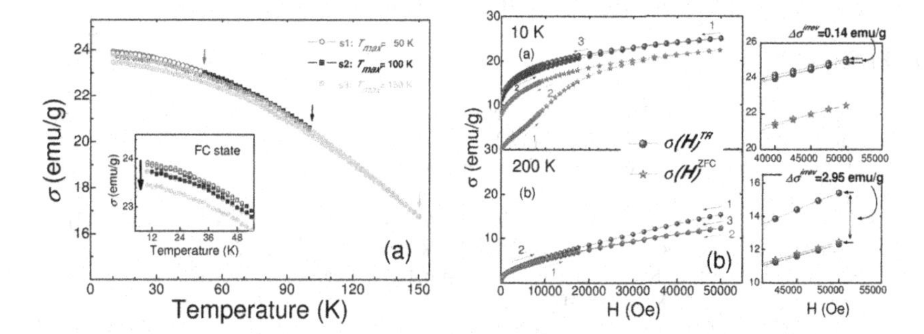

Figure 3. (a) Effect of the successive thermal cycling between 10 K and T_{max} (for T_{max} = 50 K, 100 K, and 150 K) on the $\sigma(T)$ curve at $H=50$ *kOe* after field-cooling from 350 K. The inset zooms into the low temperature region. **(b)** Demagnetization thermo-remanent curve $\sigma(H)^{TR}$ (indicated as 1) which are followed by the subsequent field-up and field-down $\sigma(H)$ curves (blue circles), measured at 10 K (a) and 200 K (b). The field-up and field-down isothermal magnetization curves measured from the demagnetized state $\sigma(H)^{ZFC}$ are also plotted (red stars). Right insets: zoom into the high-field region of the curves. The irreversible change in magnetization $\Delta\sigma(H)^{irrev}$ is indicated.

CONCLUSIONS

A kinetic arrest of the martensitic transformation has been observed in as-quenched alloy ribbons of the ferromagnetic shape memory alloy $Ni_{52.2}Mn_{34.3}In_{13.5}$ for a magnetic field above 10 kOe. The fraction of AST frozen into the martensitic matrix is not large; however, it increases with an increase of the applied magnetic field. The metastable character of the non-equilibrium field-cooled state was revealed by the decreasing behavior of magnetization after temperature cycling from 10 K to higher temperatures in the martensitic existence region. After field-cooling under a large magnetic field of 50 kOe, it was observed that if the field is totally removed the magnetization difference $\Delta\sigma$ between both states shows irreversible and reversible components, in which the irreversible component decreases with decreasing temperature. The latter contrasts with previous reports for other alloys that exhibit this abnormal phenomenon such as $Ni_{50}Mn_{34}In_{16}$ and $Mn_{50}Ni_{40}In_{10}$ alloys for which the fraction of AST frozen into the MST environment is completely metastable.

ACKNOWLEDGMENTS

Present investigation was financially supported by CONACYT, Mexico, under the projects CB-2010-01-156932, CB-2010-01-15754, and CB-2012-176705, respectively. The authors also acknowledge the support of ICyTDF, UACM, Gobierno del Distrito Federal-Mexico, and Laboratorio Nacional de Investigaciones en Nanociencias y Nanotecnología (LINAN, IPICyT). F.M. Lino-Zapata is grateful to CONACYT for supporting his Ph.D. studies. The technical support of M.Sc. G.J. Labrada-Delgado and B.A. Rivera-Escoto is gratefully recognized.

REFERENCES

1. T. Krenke, E. Duman, M. Acet, E.F. Wassermann, X. Moya, L. Mañosa, A. Planes, E. Suard, B. Ouladdiaf, Phys. Rev. B, **75**, 104414 (2007).
2. T. Krenke, E. Duman, M. Acet, E.F. Wassermann, X. Moya, L. Mañosa, A. Planes, Nature Mater. **4**, 450 (2005).
3. A.K. Pathak, M. Khan, I. Dubenko, S. Stadler, N. Ali, Appl. Phys. Lett. **90**, 262504 (2007).
4. P.A. Bhobe, K.R. Priolkar, A.K. Nigam, Appl. Phys. Lett. **91**, 242503 (2007).
5. K. Koyama, H. Okada, K. Watanabe, T. Kanomata, R. Kainuma, W. Ito, K. Oikawa, K. Ishida, Appl. Phys. Lett. **89**, 182510 (2006).
6. S.Y. Yu, Z.H. Liu, G.D. Liu, J.L. Cheng, Z.X. Cao, G.H. Wu, B. Zhang, X.X. Zhang, Appl. Phys. Lett. **89**, 162503 (2006).
7. V.K. Sharma, M.K. Chattopadhyay, K.H.B. Shaeb, A. Chouhan, S.B. Roy, Appl. Phys. Lett. **89**, 222509 (2006).
8. K. Koyama, K. Watanabe, T. Kanomata, R. Kainuma, K. Oikawa, K. Ishida, Appl. Phys. Lett. **88,** 132505 (2006).
9. T. Krenke, M. Acet, E.F. Wassermann , X. Moya, L. Mañosa , A. Planes , Phys. Rev. B **73,** 174413 (2006).
10. V.K. Sharma, M.K. Chattopadhyay, S.B. Roy, Phys. Rev. B **76,** 140401R (2007).
11. R.Y. Umetsu, W. Ito, K. Ito, K. Koyama, A. Fujita, K. Oikawa, T. Kanomata, R. Kainuma, K. Ishida, Scripta Materialia **60**, 25 (2009).
12. W. Ito, K. Ito, R.Y. Umetsu, R. Kainuma, K. Koyama, K. Watanabe, A. Fujita, K. Oikawa, K. Ishida, T. Kanomata, Appl. Phys. Lett. **92**, 021908 (2008).
13. V.K. Sharma, M.K. Chattopadhyay, S.K. Nath, K.J.S. Sokhey, R. Kumar, P. Tiwari, S.B. Roy, J. Phys.: Condens. Matter, **22**, 486007 (2010).
14. J.L. Sanchez Llamazares, B. Hernando, J.J. Suñol, C. Garcia, C.A. Ross, J. Appl. Phys. **107,** 09A956 (2010).
15. T. Sanchez, J.L. Sánchez Llamazares, B. Hernando, J.D. Santos, M.L. Sánchez, M.J. Perez, R. Sato Turtelli, R. Grössinger, Materials Science Forum, **635**, 81 (2010).
16. J.L. Sánchez Llamazares, H. Flores Zuñiga, C.F. Sánchez Valdés, C.A. Ross, C. García, J. Appl. Phys. **111**, 07A932 (2012).

Mater. Res. Soc. Symp. Proc. Vol. 1485 © 2013 Materials Research Society
DOI: 10.1557/opl.2013.285

Thermal Properties of Cu-Hf-Ti Metallic Glass Compositions

I. A. Figueroa

Instituto de Investigaciones en Materiales, Universidad Nacional Autónoma de México, Cd. Universitaria, Del. Coyoacán, México D.F. C. P. 04510, México.

ABSTRACT

The glass transition temperature T_g, crystallization temperature T_x, solidus temperature T_m, and liquidus temperature T_l, of a number of ternary Cu-Hf-Ti glassy alloys in the composition range of 51< Cu <67, 5 < Hf < 40 and 5 < Ti <40 (at.%) are reported and discussed. It is found that increasing the Ti:Hf ratio results in a rapid decreasing of T_g and T_x. This behavior is related to the fact that the melting point and cohesive energy for Ti are substantially lower than for Hf. The solidus temperature T_m, remains relatively constant on a wide range of compositions. The *liquidus* temperatures data suggest a ternary eutectic within the compositional field encompassed by the $Cu_{55}Hf_{20}Ti_{25}$, $Cu_{59}Hf_{21}Ti_{20}$, $Cu_{60}Hf_{20}Ti_{20}$ and $Cu_{55}Hf_{21}Ti_{24}$ alloys, with a liquidus temperature, T_l, of ~1170 K; this is supported by the DTA traces, which show a single melting peak. Based on the DTA analysis, the experimentally calculated *liquidus* projection for the ternary Cu-Hf-Ti alloy system is also reported.

INTRODUCTION

Metallic glasses have gained considerable interest due to their unique properties. These key alloys have been studied because of their wide range of applications and exceptional properties, e.g., high thermal stability, good glass forming ability (GFA), high mechanical strength, soft magnetic properties and excellent corrosion resistance [1]. The systems including alloys based on Cu, Mg, Zr , Ti , Fe, Pd–Cu, Co and Ni [2-9], have been the most investigated. Thermal properties are also very important to determine the maximum working temperature of the metallic glass and its capability of forming a glassy phase. To date the most important thermal properties in metallic glasses have been the glass transition temperature T_g, crystallization temperature T_x, *solidus* temperature T_m, and *liquidus* temperature T_l). Recently, some thermal properties in the neighborhood of the alloy $Cu_{60}Hf_{25}Ti_{15}$ system are examined in the composition range of 57.5<Cu *at.%*<62.5, 22.5<Hf *at.%*<28.75 and 11.25<Ti *at.%*<17.5. These alloys show values of $T_g \sim 723\ K$, $T_l \geq 1188\ K$, $T_x \sim 773\ K$ and $\Delta T \sim 50\ K$. In previous works, the present author has found values of $T_g \sim 743\ K$ and $T_x \sim 773\ K$ on $Cu_{55}Hf_{30}Ti_{15}$ and $Cu_{55}Hf_{25}Ti_{20}$ systems and values of $T_g \sim 750\ K$ and $T_x \sim 800\ K$ on $Cu_{65}Hf_{30}Ti_{10}$ and $Cu_{55}Hf_{40}Ti_5$ alloy systems [10-11]. The aim of this work is to carry out a detailed analysis of the thermal properties such as T_g, T_x, T_m and T_l, of Cu-Hf-Ti glassy alloy ribbons, produced by melt spinner, in the composition range of 51< Cu <67, 5 < Hf < 40 and 5 < Ti < 40 (at.%), and to determinate its behavior as a function of the Cu, Hf and Ti content.

EXPERIMENTAL DETAILS

$Cu_xHf_yTi_z$ alloys, where x = 51–67 *at.%*, y = 5–40 *at.%* and z = 5–40 *at.%*, are prepared by argon arc melting mixtures of Cu (99.99 % pure), Hf (99.8 % pure) and Ti (99.6 % pure). The alloy compositions represent the nominal values but the weight losses in melting are negligible (< 0.1 %). The alloy ingots are inverted on the hearth and re-melted several times to ensure compositional homogeneity. Ribbon samples of each alloy are prepared by means chill block melt-spinning technique in a sealed helium (He) atmosphere at roll speeds in the range of 3 *m/s* to 35 *m/s*. The magnitude of T_g and T_x are determined by DSC (using a Perkin Elmer DSC-7). The T_m and T_l are measured by DTA (using Perkin Elmer DTA-7). Both experimental procedures are realized to heating rates of 0.33 *K/s*. To confirm the reproducibility of the results, at least three samples are measured for each composition.

DISCUSSION

The ribbon samples thicknesses are in the range 18 *μm* to 180 *μm*, which correspond to roll speed range of 3 *m/s* - 35 *m/s*. XRD patterns of melt spun ribbons cast obtained for the ternary $Cu_xHf_yTi_z$ system (x = 51–67 *at.%*, y = 5–40 *at.%* and z = 5–40 *at.%*) consist only of a broad diffraction maxima (30º–40º) without any sharp Bragg peaks, indicating that the ribbons has a fully glassy structure.

Figure 1a shows the DSC scans for some of the investigated alloys. In all cases, a clear endothermic event associated with the glass transition is evident, followed by an undercooled liquid region prior to the exothermic crystallization event. Figure 1b and 1c show the compositional maps for the glass transition (T_g) and crystallization (T_x) temperatures for the Cu-Hf-Ti alloy system. It can be seen that increasing the Ti:Hf ratio results in rapid decreases in T_g and T_x, being associated to the fact that the melting point and cohesive energy for Ti are substantially lower than for Hf. Meanwhile, Figure 2 shows an isoplethal section of the Cu-Hf-Ti ternary alloy system, where the variation in T_g and T_x with the Ti content for $Cu_{55}Hf_{45-x}Ti_x$ glassy alloys can be appreciated more clearly. The magnitude of T_g decreases monotonically between x = 5 and 40 from 750 *K* to 670 *K*. Similarly, T_x decreases monotonically from 800 *K* at x = 5 to 695 *K* at x = 40. This behavior has been reported previously in $Cu_{60}Hf_{40-x}Ti_x$ alloys with the same Ti content [2], where T_g and T_x also decrease progressively with increasing Ti content up to x = 35. This corresponds to a change in the crystallization mode from a single peak to two or three stages.

The *solidus* T_m and *liquidus* T_l isotherms based on the results of the DTA analysis for some investigated ternary Cu-Hf-Ti alloys are plotted in Figure 3a and 3b. Although there does not appear to be a close resemblance between the *liquidus* projection for the ternary Cu-Hf-Ti alloy system shown in Figure 3b with the *liquidus* projection calculated from thermodynamic data for the Cu–Ti–Zr system [12], the overall the ternary eutectic compositions can be interpreted as being broadly similar. Certainly, further experimental work is needed to obtain a complete picture from the ternary Cu-Hf-Ti system, as thermodynamic data for the Cu-Hf-Ti ternary system is not available in the literature.

Nevertheless, the T_l data obtained from the present study gives a good idea of the eutectic point of this system, with T_l being just below 1170 *K* at a composition such as $Cu_{55}Hf_{25}Ti_{20}$. Figure 2 also shows the trend of T_m as a function of Ti content; it decreases from 1193 *K* to 1143 *K* for $Cu_{55}Hf_{40}Ti_5$ and $Cu_{55}Hf_{30}Ti_{15}$, respectively, and then, remains fairly constant at ~ 1140 ± 10 *K* with further increases of Ti up to 40 *at.%*.

As mentioned above, the T_l data suggest a ternary eutectic within the compositional field encompassed by the $Cu_{55}Hf_{20}Ti_{25}$, $Cu_{59}Hf_{21}Ti_{20}$, $Cu_{60}Hf_{20}Ti_{20}$ and $Cu_{55}Hf_{21}Ti_{24}$ alloys, with a T_l of ~ 1170 *K*; this is supported by the DTA traces which show a single melting peak, for these compositions, as shown in Figure 4.

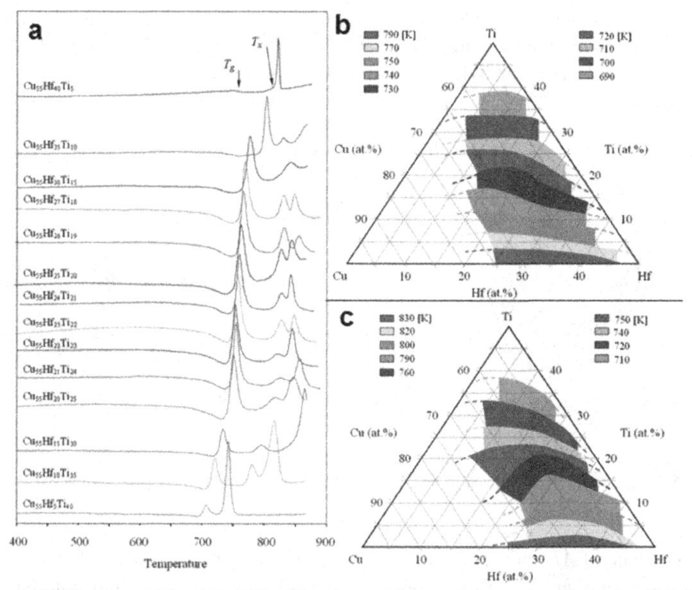

Figure 1. a) DSC traces of the $Cu_{55}Hf_{45-x}Ti_x$ glassy alloy series at a heating rate of 0.33 *K/s,* b) Compositional map for the glass transition temperatures (T_g) and c) Compositional map for the crystallization temperatures (T_x) covering the area of the Cu-Hf-Ti system investigated.

Figure 2 also shows clearly the variation of T_l with the Ti content for $Cu_{55}Hf_{45-x}Ti_x$ glassy alloys, where T_l decreases to a minimum value of 1168 *K* at $x \sim 25$ and then increases up to 1213 *K* at $x \sim 45$. It is worth mentioning that the glass forming ability (GFA) is commonly evaluated by different thermally obtained parameters such as T_{rg}, ΔT_x, ΔT_l, α, δ, ϕ, γ, ω, β and β' [1, 13-20]. Since these parameters could be easily calculated from the results reported in this work and their predicting capability is fairly the same, only some of these parameters will be given here.

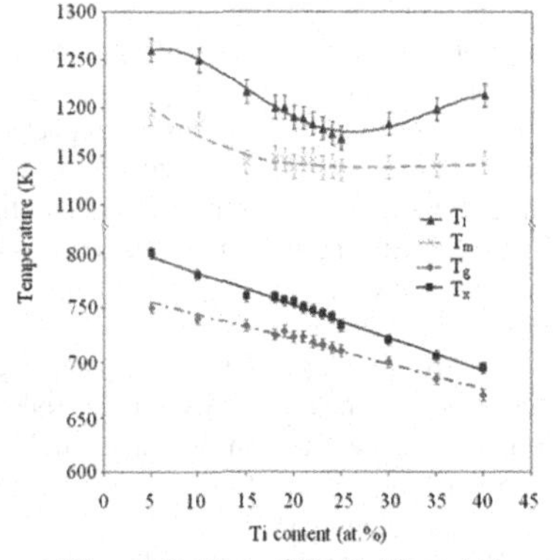

Figure 2. Plot of T_g, T_x, T_m and T_l as a function of Ti content for the $Cu_{55}Hf_{45-x}Ti_x$ alloy series.

Figure 5a displays the compositional map of T_{rg} for the range of Cu-Hf-Ti alloys investigated. The values of T_{rg} ($T_{rg} = T_g/T_l$) and others, are calculated in order to report where the highest GFA could be located. The highest values of T_{rg} are found in the $Cu_{56}Hf_{25}Ti_{19}$, $Cu_{59}Hf_{21}Ti_{20}$ and $Cu_{20}Hf_{20}Ti_{20}$ alloys with T_{rg} ranging from 0.61 to 0.62. Consequently, the area with the highest T_{rg} value of ~ 0.62 is located within a compositional range of 55–60 at.% Cu, 20–25 at.% Hf, and 18–21 at.% Ti. The parameter ΔT_x $(=T_x-T_g)$, showed an overall value of ~ 30 K for most of the alloys and only some alloys with low Ti at.% had large values of ΔT_x, e.g. $Cu_{60}Hf_{35}Ti_5$, $Cu_{55}Hf_{45}Ti_5$ and $Cu_{62}Hf_{25}Ti_{13}$ with 57K, 55K and 50 K, respectively.

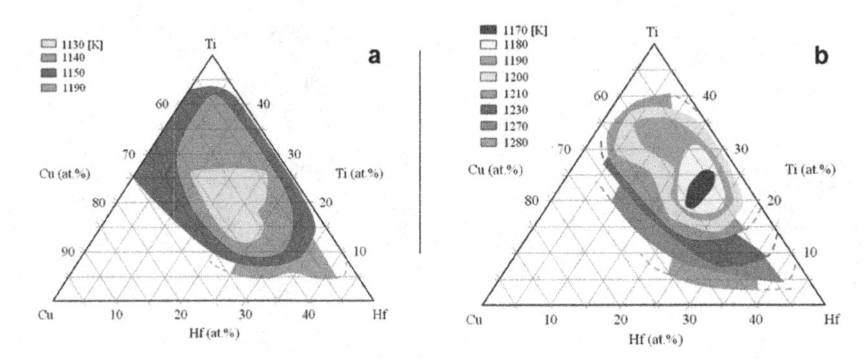

Figure 3. a) *Solidus* projection for the Cu–Hf–Ti system and b) *Liquidus* projection for the Cu–Hf–Ti system.

A large value of ΔT_x indicates that an alloy can be heated substantially above T_g without inducing devitrification and thus that the alloy in a highly undercooled liquid state has a high resistance to crystallization. In this study, ΔT_x, had maxima at low Ti content, where it is thought that the oxygen content is also low, followed by a decrease with further increments in Ti content, i.e. high oxygen content.

Figure 5b and 5c show the maps of γ and δ parameters, respectively. The highest values of γ and δ parameter are found in the same $Cu_{60}Hf_{20}Ti_{20}$, $Cu_{59}Hf_{21}Ti_{20}$ and $Cu_{56}Hf_{25}Ti_{19}$ alloys, with 0.40 and 1.7 respectively. The area with the highest T_{rg}, γ and δ parameters is located within a compositional range of 55–60 *at.%* Cu, 20–25 *at.%* Hf, and 18–21 *at.%* Ti. Figure 5d shows the compositional map of the magnitude of undercooling required to form an amorphous solid, ΔT_l (T_l-T_g). The area with the lowest ΔT_l value of ~ 450 K is located within the same compositional range as the maximum in T_{rg} (55–60 at.% Cu, 20–25 at.% Hf, and 18–21 at.% Ti). The $Cu_{59}Hf_{21}Ti_{20}$ and $Cu_{60}Hf_{20}Ti_{20}$ alloy compositions had the smallest undercooling value of 448 K of all the alloys investigated in this system. Figure 5b and 5c show the maps of γ and δ parameters, respectively. The highest values of γ and δ parameters, respectively. The highest values of γ and δ parameter are found in the same $Cu_{60}Hf_{20}Ti_{20}$, $Cu_{59}Hf_{21}Ti_{20}$ and $Cu_{56}Hf_{25}Ti_{19}$ alloys, with 0.40 and 1.7 respectively. The area with the highest T_{rg}, γ and δ parameters is located within a compositional range of 55–60 *at.%* Cu, 20–25 *at.%* Hf, and 18–21 *at.%* Ti. Figure 5d shows the compositional map of the magnitude of undercooling required to form an amorphous solid, ΔT_l (T_l-T_g). The area with the lowest ΔT_l value of ~ 450 K is located within the same compositional range as the maximum in T_{rg} (55–60 at.% Cu, 20–25 at.% Hf, and 18–21 at.% Ti). The $Cu_{59}Hf_{21}Ti_{20}$ and $Cu_{60}Hf_{20}Ti_{20}$ alloy compositions had the smallest undercooling value of 448 K of all the alloys investigated in this system. This is consistent with the location of the eutectic (Figure 3b) and with the value of T_{rg}, γ and δ obtained for same alloys.

The smaller the undercooling required, the lower the driving force for crystallization and the longer the time for any nucleants, both heterogeneous and homogenous in nature, to grow and form crystals (T_{rg} does not take into account nucleation growth kinetics or extrinsic factors such as heterogeneous nucleation, but it does take account of homogeneous nucleation, but in a very empirical way). Therefore, the closer the freezing point of an alloy is to T_g, the better the GFA should be, i.e. this should be greater for a eutectic or near-eutectic alloy. Thus, for an alloy composition to maximize the possibility of forming a glassy phase, the value of T_l-T_g should be normally minimized, notwithstanding any unusual concentration dependence of the number density and/or effectiveness of heterogeneous nucleants.

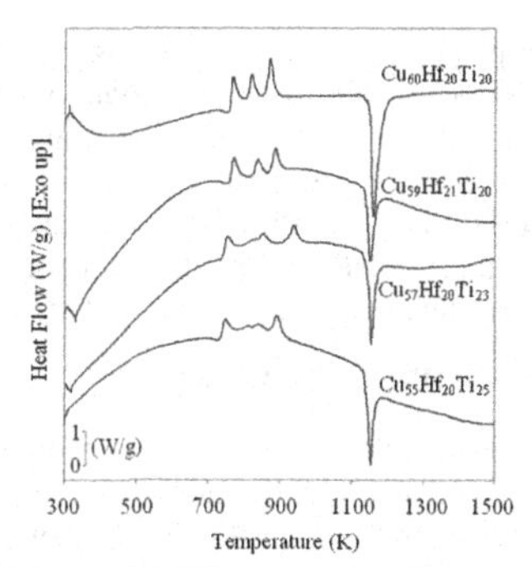

Figure 4. DTA curves for glassy eutectic alloys in the Cu-Hf-Ti system, heating rate of 0.33 K/s.

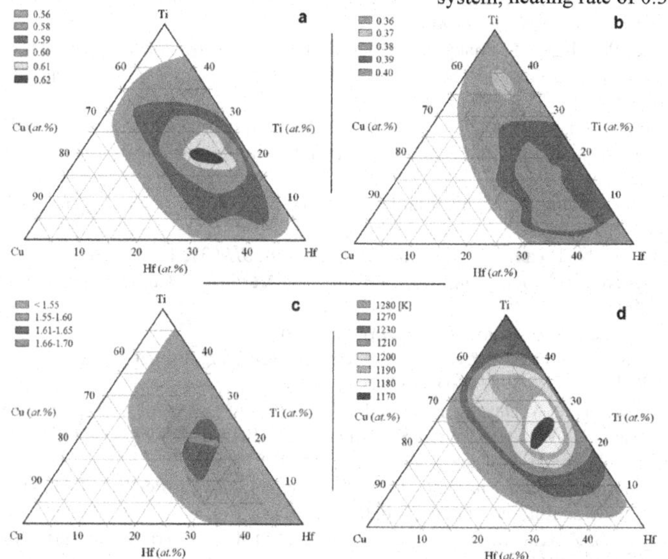

Figure 5. Compositional maps for a) T_{rg}, b) γ , c) δ and d) ΔT_l, for Cu–Hf–Ti alloys investigated.

CONCLUSIONS

The thermal properties a number of compositions in the Cu-Hf-Ti alloy family are investigated. The magnitude of T_g and T_x decreased progressively when increasing Ti. The *solidus* temperature T_m, remained relatively constant on a wide range of compositions with an average value of 1130 ± 10 K. The *liquidus* temperatures data suggested a ternary eutectic within the compositional field encompassed by the $Cu_{55}Hf_{20}Ti_{25}$, $Cu_{59}Hf_{21}Ti_{20}$, $Cu_{60}Hf_{20}Ti_{20}$ and $Cu_{55}Hf_{21}Ti_{24}$ alloys, with a *liquidus* temperature, T_l, of ~1170 K; this is supported by the DTA traces which showed a single melting peak, for these compositions. Based on the results from the DTA analysis, the first the experimental *liquidus* projection, reported in the literature, for the ternary Cu-Hf-Ti alloy system is reported in this work. According to the different thermally obtained parameters for predicting the glass forming ability (GFA), the best glass forming compositions could be located near or at the eutectic.

ACKNOWLEDGMENTS

IAF acknowledges the financial support of PAPIIT-UNAM "IB100712". A. Tejeda, C. Flores, G. Aramburo, J. J. Camacho, G. Lara Rodriguez, R. Reyes, O. Novelo, J. Arellano, J. Morales-Rosales and C. Gonzalez-Sanchez are also acknowledged for their technical support.

REFERENCES

1. Y. J. Yang, D. W. Xing, J. Shen, J. F. Sun, S. D. Wei, H. J. He and D. G. McCartney, *J. Alloys Compd.* 415, 106 (2006).
2. A. Inoue, W. Zhang, T. Zhang and K. Kurosaka, *Acta Mater.* 49, 2645 (2001).
3. I. A. Figueroa, R. Rawal, P. Stewart, P. A. Carroll, H. A. Davies, I. Todd and H. Jones, *J. Non-Cryst. Solids* 353, 839 (2007).
4. I. A. Figueroa, H. A. Davies, I. Todd and K. *Yamada, Adv. Eng. Mater.*, 9 (2007) 496A
5. A. Inoue, A. Kato, T. Zhang, S.G. Kim and T. Masumoto, *Mater. Trans. JIM,* 32, 609 (1991).
6. A. Inoue, T. Zhang, N. Nishiyama, K. Ohba and T. Msumoto, *Matter. Lett.* 19, 131 (1994).
7. A. Inoue, N. Nishiyama, T. Matsuda, *Mater. Trans. JIM* 37, 181 (1996).
8. I. A. Figueroa, J. I. Betancourt, G. Lara and J. A. Verduzco, *J. of Non-Cryst. Solids,* 329, 3075 (2005).
9. T. Itoi and A. Inoue. *Mater. Trans. JIM* 42, 1256 (2000).
10. L.C. Damonte, A.F. Pasquevich, L.A. Mendoza-Zelis, I.A. Figueroa Vargas, H.A. Davies and I. Todd, *Phys. B* 398, 480 (2007).
11. I. A. Figueroa, H. A. Davies and I. Todd, *J. Alloys and Comp.* 434–435, 164 (2007).
12. R. Arroyave , T. W. Eagar and L. Kaufman, *J. of Alloys and Comp.* 351, 158 (2003).
13. I. W. Donald and H. A. Davies, Phil. Mag. 42, 277 (1980).
14. I A. Figueroa, S. Baez-Pimiento, J. D. Plummer, O. Novelo-Peralta, H. A. Davies and I. Todd, Acta Metallurgica Sinica, 25-6, 0409 (2012).
15. G. Q. Liu, S. Z. Kou, C. Y. Li, Y. C. Zhao and H. L.Suo, Trans. Nonferrous Met. Soc. China 22, 590 (2012).
16. P. Gong, K. F. Yao and Y. Shao, Journal of Alloys and Compounds 536, 26 (2012).
17. B. S. Dong, S. X. Zhou, D. R. Li, C. W. Lu, F. Guo, X. J. Ni and Z. C. Lu, Prog. Nat. Sci.: Mater. Inter. 21, 164 (2011).
18. Z. L. Long, H. Q. Wei, Y. H. Ding, et al. J Alloys Compd 475, 207 (2009).
19. X. Ji and Y. Pan, Trans. Nonferrous Met. Soc. China 19, 1271 (2009).
20. C. Suryanarayana, I. Seki and A. Inoue, J Non-Cryst Solids 355, 355 (2009).

Mater. Res. Soc. Symp. Proc. Vol. 1485 © 2013 Materials Research Society
DOI: 10.1557/opl.2013.286

On the characterization of eutectic grain growth during solidification

M. Morua, M. Ramirez-Argaez and C. Gonzalez-Rivera C.
Departamento de Ingeniería Metalúrgica, Facultad de Química, UNAM, 04510, México D.F.

ABSTRACT

The purpose of this work is to compare the results obtained from three methodologies intended to estimate kinetic parameters describing quantitatively the grain growth during equiaxed eutectic solidification in order to identify the best procedure to characterize grain growth kinetics. A heat transfer / solidification kinetics model is implemented to simulate the cooling and solidification of eutectic Al-Si and eutectic cast iron in sand molds. Using simulated cooling curves and volume grain density data generated by the model, the three methods are applied to obtain their predicted grain growth coefficients. The predicted results are compared with the grain growth coefficients used in the model. The outcome of this work suggests that two of the three methods under study represent the best option to obtain the kinetic parameters of equiaxed growth during eutectic solidification.

INTRODUCTION

The simulation of the cooling and solidification of alloys has been an important tool to improve metal processing productivity in the last decades. The prediction of the formation and evolution of the solidification microstructure is intimately linked to understanding of solidification kinetics. In this regard, different modeling approaches have been developed in the last couple of decades. A recent work by Nakajima et al. [1] discusses the methodological progress for computer simulation of solidification.

The models used in computer simulations of solidification involve calculations on nucleation and growth kinetics which in turn depend on the availability of laws of nucleation and growth capable to reproduce the experimental behavior of the alloys of interest.

The simulation software commonly available includes and uses material data for some generic alloys with compositions according to standard specifications. However a lack of data in existing databases regarding more specific alloy systems could be a restriction to the prediction capability of this computational tool. For this reason, identification of the more suitable methods to obtain experimental data to simulate the solidification and grain growth kinetics of specific alloys is an important task.

Computer aided cooling curve analysis methods (CA-CCA) have been used to study solidification kinetics of various alloy systems of metallurgical interest. It has been found that Fourier thermal analysis (FTA) [2] is the most reliable CA-CCA method because it takes into account the presence of thermal gradients, giving a more realistic evolution of the solid fraction.

FTA method has been described in detail elsewhere [2]. Briefly, two cooling curves obtained at two radial locations within a cylindrical sample with thermally isolated top and bottom during its cooling and solidification, are numerically processed. The FTA numerical processing starts with the generation of the first derivative with respect to time of the curve corresponding to the thermal center of the sample and the identification of the times of start and end of solidification. FTA uses the data acquired from the two thermocouples to obtain the thermal diffusivity of the sample and the zero baseline curve by an iterative procedure. The integration of the area between the first derivative of the cooling curve and the zero baseline curve gives quantitative data on the evolution of solid fraction, fs, and solidification rate dfs/dt. Some methodologies allowing grain growth characterization have been developed in order to obtain more detailed information about kinetics of eutectic equiaxed growth.

It has been found that, during solidification of eutectic binary alloys, eutectic equiaxed grain growth rate depends on the undercooling, ΔT, defined as the difference between the equilibrium solidification temperature and the actual temperature present during the phase change, Eq. (1)

$$\Delta T = T_{Eu} - T \tag{1}$$

In Eq.(1) T_{Eu} is the equilibrium eutectic temperature and T is the instantaneous temperature recorded by the thermocouple during solidification in undercooled melts.

The following exponential equation has been used to describe equiaxed grain growth, where μ and n are the eutectic growth coefficients, R is the grain radius, t is time, and the derivative of R with respect to time is the grain growth rate..

$$\frac{dR}{dT} = \mu \Delta T^{n} \tag{2}$$

In the method proposed by Degand et al. [3], it is assumed that n = 2, according to the classical eutectic growth theory, and the pre exponential parameter μ of Eq. (2) is obtained from experimental data using the following equation:

$$\mu = \frac{R_{Av}}{\int_{t_{min}}^{t_{0.74}} \Delta T^{2} dT} \tag{3}$$

In Eq.(3) R_{Av} is the mean grain radius, ΔT is the instantaneous undercooling defined in Eq. (1), t_{min} is the time corresponding to the maximum undercooling after the start of solidification and $t_{0.74}$ is the time to reach a solid fraction of 0.74, which is assumed to indicate the impingement of growing spherical grains. R_{Av} is obtained using standard metallographic etching methods. The integral in Eq. (3) is solved numerically by using instantaneous undercooling data and t_{ini} determined from the cooling curve. The time of grain growth impingement t_{imp} is identified from data of solid fraction evolution obtained from CA-CCA methods.

There are two other methods used to obtain the eutectic growth coefficients. Both methods assume that, during solidification of the sample, there are N spherical grains of mean radius R growing simultaneously. The main difference between the two methods is the grain growth model that is used during calculations.

In the method of Dioszegi and Svensson [4], the grain growth model is the Kolmogorov, Johnson, Mehl and Avrami model that takes into account grain impingement in the last stages of solidification. The solid fraction at time t is given by Eq. (4) where fs^{t} and R^{t} are the solid fraction and grain radius at time t respectively:

$$f_s^t = 1 - \exp(-\tfrac{4}{3}\pi N (R^t)^3) \quad (4)$$

For the Free Grain Growth method [5], a section of the cooling curve, associated to the first stages of solidification, is used (i.e. grain impingement is neglected). It is assumed that during solidification, N spherical grains of mean instantaneous radius R^t, develop freely at the same time. Thus the solid fraction in the early stages of growth at a time t is given by Eq. (5):

$$f_s^t = \tfrac{4}{3}\pi N (R^t)^3 \quad (5)$$

If the density of grains per unit volume in the sample, N, can be known from metallographic characterization, and fs^t is obtained using FTA, then it is possible to estimate the instantaneous grain radius R^t for both grain growth models. For both methods, the evolution of grain radius as a function of time (i.e. the grain growth rate), can be obtained by numerical differentiation:

$$\frac{dR}{dT} = \frac{R^{t+dt} - R^t}{\Delta t} \quad (6)$$

The purpose of this work is to compare results obtained from the three described methods and identify the best procedure to characterize grain growth kinetics.. In order to reach this goal, a heat transfer solidification kinetics model is implemented to simulate the cooling and solidification of eutectic Al-Si and eutectic cast iron in sand molds. Using simulated cooling curves and grain density data generated by the model the three methods are applied to obtain their predicted grain growth coefficients. Finally the results predicted by each method are compared with the grain growth coefficients used in the model.

HEAT TRANSFER AND SOLIDIFICATION KINETICS MODEL

A heat transfer solidification kinetics (HT/SK) model, previously validated with experimental results [6], is used to simulate the cooling and solidification of eutectic Al-Si in sand molds in order to generate the cooling curves and grain density information required to evaluate the grain growth coefficients. Additionally, this model is modified to allow the simulation for eutectic cast iron solidification by using the thermophysical properties and nucleation model for eutectic cast iron reported by Maijer et al. [7], and the grain growth coefficients reported by Dioszegi and Svensson [4].

Heat transfer.

During cooling and solidification of a cylindrical casting in a sand mold, it is assumed that the macroscopic heat flow is governed by conductive heat transfer and latent heat generation due to solidification. The model assumes that heat transfer occurs only by unidirectional radial conduction in cylindrical coordinates. Constant thermo physical properties are also assumed. Thus the energy balance applied to the metal/mold system can be written as:

$$Cp_j^v \frac{\partial T(r,t)}{\partial t} = k_j^{th} \frac{1}{r}\frac{\partial}{\partial r}\left(r \frac{\partial T(r,t)}{\partial r}\right) + L_f^v \frac{\partial f_s(r,t)}{dt} \quad (7)$$

where Cp_j^v is the volumetric heat capacity and the subscript j indicates the metal domain (j=1) or the mold domain (j=2), T(r, t) is the temperature, r is the radial position, t is the time, k^{th} is the thermal conductivity and L_f^V is the volumetric heat of fusion

Solidification Kinetics.

The eutectic grains are assumed to be spherical in shape. The eutectic grains growth rate is calculated by using Eq. (2). Table 1 shows the growth parameters used for calculation in the cases of eutectic Al-Si [6] and eutectic cast iron [4] materials.

Table 1. Eutectic Growth coefficients used during simulation

Parameter	Eutectic Al-Si	Eutectic cast iron	Units
μ	5×10^{-6}	4.8×10^{-7}	$m\ S^{-1}\ °C^{-2}$
n	2	0.66	-

Eq. (7) is solved by generating a discretized description of the cylindrical metal/mold system in the form of a finite difference mesh composed by a known number of cylindrical volume elements, and using the implicit finite difference method. During solidification, the local changes in the solid fraction within the casting are calculated at the beginning of every interval of time by inputting the temperature field at this time into the solidification kinetic model. The results obtained are then used during the calculation of the next temperature field. The assumptions, including the initial and the boundary conditions used to solve Eq.(7) are described elsewhere [6].

During simulation, the radius of the casting is fixed at r = 0.015 m. Two cooling curves are obtained for volume elements located at two different radial positions (r_1= 0 and r_2 = 0.005 m). The cooling curves are simulated by the model and numerically processed using FTA method. The volume grain densities predicted by the model are 5.2×10^8 grains/m^3 and 1.5×10^{10} grains/m^3 for eutectic Al-Si and eutectic cast iron respectively.

RESULTS AND DISCUSSION

Fig. 1 shows the cooling curves predicted by the model and the solid fraction evolution as obtained from FTA numerical processing of these curves for the two eutectics under study.

The integral shown in eq.(3) to obtain μ using the method of Degand is calculated with the cooling curve data to obtain undercooling ΔT from the time of maximum undercooling to the time of impingement of grains, shown in Fig. 1 as t_{min} and $t_{0.74}$ respectively.

On the other hand, for the methods of Dioszegi and Free Growth, the evolution of the grain growth rate as a function of undercooling is calculated using the known number of grains per unit volume, N, Eqs. (1) and (6) and the data of the cooling and solid fraction curves from t_{min} to t_{max}, the time of maximum recalescence, also shown schematically on Fig.1.

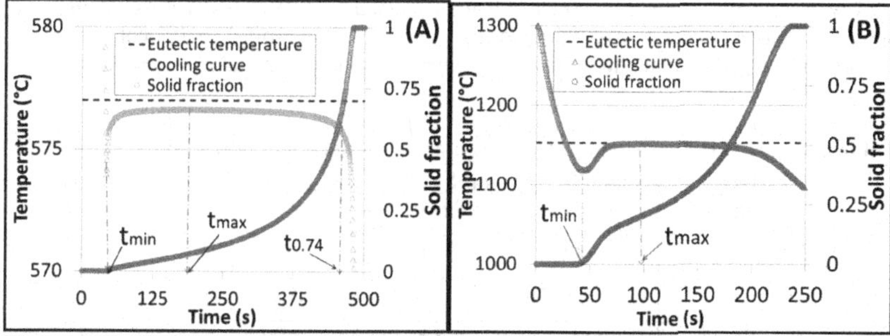

Figure 1. Cooling curves predicted by the HT/SK eutectic model and FTA solid fraction evolution and description of the time interval used to perform grain growth characterization for (A) eutectic Al-Si and (B)eutectic cast iron.

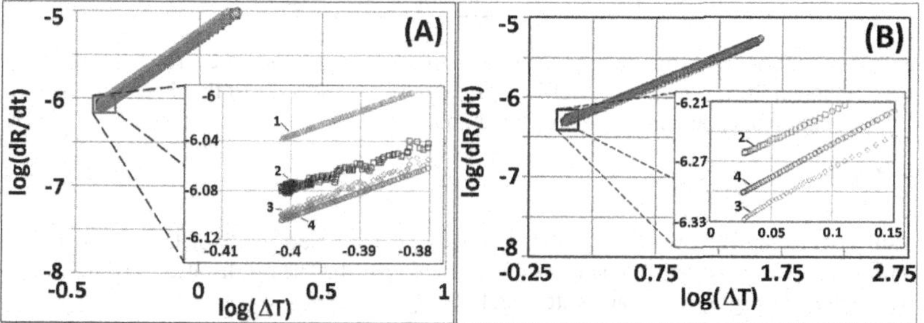

Figure 2. Logarithmic plot of grain growth rate against undercooling obtained from the methods: 1.Degand, 2.Dioszegi, 3. Free Growth and 4. HT-SK model applied to eutectic Al-Si (A) and eutectic cast iron (B) alloys.

Fig. 2 shows the grain growth rate, dR/dt as a function of undercooling, ΔT obtained from the methods under consideration in this work. The graph is constructed using a logarithmic plot. In all cases is found a linear tendency. Fig. 2(A) corresponds to the Al-Si eutectic solidification. The inset of this figure shows that Degand's method predicts a larger difference as compared to the HT-SK model data, and this is followed by Dioszegi's and Free Growth methods.

Fig. 2(B) shows predictions of Dioszegi's and Free Growth methods and the data generated by the HT-SK model for eutectic cast iron solidification. These methods can be used to determine the exponential coefficient **n** in Eq.(2). The corresponding inset shows that Dioszegi's method predicts the larger difference when compared to the HT-SK model data. The Free Growth method produces slightly better predictions for the growth coefficients.

Table 2 presents numerical results concerning the eutectic growth coefficients as determined for the cases under study, μ_{Ref} and n_{Ref} are the values of the growth parameters used by the HT-SK models. Apparently the methods of Dioszegi and Free Growth are better suited than the method of Degand to obtain growth coefficients for Al-Si eutectic solidification. The effect of impingement is important after the first stages of solidification and the method of Degand uses undercooling data corresponding to relatively high solid fractions without taking into account this effect. This feature could explain the observed results.

However, Degand's method is still capable to generate a relatively good approximation, around at least the correct order of magnitude, for the pre exponential growth coefficient of the alloy. Thus it can be a good method to obtain a first approximation of μ by taking into account its simplicity.

Table 2. Growth kinetics parameters obtained using the methods under study.

Eutectic	Method	**μ_{Ref}=5x10^{-6}**	**n_{Ref}=2**
Al-Si	Degand	5.82x10^{-6}	----
Al-Si	Dioszegi	4.91x10^{-6}	1.92
Al-Si	Free growth	5.01x10^{-6}	1.99
Eutectic	Method	**μ_{Ref} = 4.8x10^{-7}**	**n_{Ref} = 0.66**
Cast iron	Dioszegi	5.17 x10^{-7}	0.64
Cast iron	Free growth	4.49 x10^{-7}	0.67

In the case of eutectic cast iron, results show that both Dioszegi's and Free Growth methods could be used to obtain the grain growth coefficients. The results also suggest that Free Growth method is slightly more accurate than Dioszegi's method.

CONCLUSIONS

The results obtained in this work suggest that Dioszegi´s and Free Growth methods represent the best option to obtain the kinetic parameters of equiaxed growth. The results also show that Free Growth method is slightly more accurate than Dioszegi's method.

Degand's method is capable to predict approximately the pre exponential growth coefficient and it could be a good method to obtain a first approximation of this parameter.

ACKNOWLEDGMENTS

The authors acknowledge DGAPA UNAM, for its financial support (Project IN113912) and to A. Ruiz, and A. Amaro for their valuable technical assistance. Begin typing text here.

REFERENCES

1. K.Nakajima,H. Zhang, K.Oikawa, M. Ohno and P.G. Jonsson. *ISIJ Int*, 50, 1724 (2010)
2. E.Fras .F. Kapturkiewicz, A. Burbielko, H.F. Lopez, *AFS Trans.*, 101, 505(1993),
3. C.Degand, D.M. Stefanescu, and G. Laslaz, Proc. *Int. Symp. On Advanced Materials and Technology for the 21st Century*, Honolulu,USA, 55, (1995).
4. A. Dioszegi and I. L. Svensson. *Int.J. of Cast Metal Res.,* 18, 41(2005).
5. M. Morua, M. Ramirez-Argaez and C. Gonzalez-Rivera. Experimental determination of grain growth kinetics during eutectic solidification, *Mater. Science: An Indian Journal,* Article in press, (2012).
6. C. Gonzalez-Rivera, B. Campillo, M. Castro, M. Herrera, J. Juarez., *Mater. Science and Eng. A.*, A279, 149(2000).
7. D. Maijer, S.L. Cockcroft and W. Patt. *Metall. Mater. Trans A*, 30A, 2147(1999).

AUTHOR INDEX

Alfaro-López, E., 29
Altamirano, G., 83, 143
Alvarado-Tenorio, B., 77, 137
Amaro-Villeda, Adrián M., 101
Atlatenco, E. Cándido, 113
Ávila-Ambriz, J.L., 53

Baltazar-Hernandez, V.H., 95
Basso, Alejandro D., 113, 125
Bedolla-Jacuinde, A., 143
Bykov, A., 65

Cabrera, J.M., 1, 83, 143
Cadenas-Calderón, E., 53
Calderon, H.A., 71
Campillo, B., 139
Castro-Román, M.J., 29
Clark, Braeden M., 9
Conejo, A., 101

de Landa Castillo-Alvarado, Fray, 21
Domínguez-Almaraz, G.M., 53

Figueroa, I.A., 155
Flores, O., 137

García-Fernández, T., 149
García-Hernández, Claudia M., 89
García, José A., 113, 125
García-Mora, E., 143
García-Pastor, Francisco Alfredo, 29, 35
Garza-García, Mitzué, 47
Gleń, Marta A., 59
Gonzalez-Rivera, C., 161
Graeve, Olivia A., 9
Grzmil, Barbara U., 59
Gutiérrez, Emmanuel J., 119

Haro-Rodriguez, S., 95
Hernández-Expósito, A., 83
Hernández-Hernández, M., 41
Hernández-Ibarra, Oscar, 47
Huerta-Ricardo, A., 71

Jaffe, M., 137

Kakazey, M., 65
Kelly, James P., 9

Lara-Rodríguez, A.G., 149
Lino-Zapata, F.M., 149
López-Cuevas, Jorge, 47, 89, 107
López-García, R.D., 29
López-Ibarra, A., 95
López, N.M., 131

Márquez-Aguilar, P.A., 65
Martínez-González, C.J., 95
Mejía, I., 83, 143
Morua, M., 161

Nayak, S.S., 95

Parada-Soria, A., 77
Peyraut, F., 53

Ragulya, A., 65
Ramírez-Argáez, M.A., 41, 101, 161
Ramírez-Rodríguez, Teresa, 21
Ramos-Gómez, E.A., 41
Ramos-Ramírez, Magaly V., 107
Reyes-Mayer, A., 137
Ríos-Jara, D., 149
Rivero, H.D., 113
Rodríguez-Galicia, José L., 89, 107
Romo-Uribe, A., 77, 137
Ruiz, Gerardo A., 125

Salinas, Armando, 119
Salinas R.A., 131
Sanchez-Cadena, L., 77
Sánchez Llamazares, J.L., 149
Sicora, Jorge, 113, 125
Solórzano-López, Juan, 35

Stetsenko, V., 65

Tsuchiya, K., 71

Umemoto, T., 71

Varela-Castro, G., 1
Vlasova, M., 65

Yao, HF, 77

Zenil, Ramses, 125
Zhou, Y., 95

SUBJECT INDEX

additives, 59
al, 41, 53
alloy, 113

B, 83

C, 9, 21
ceramic, 47, 65, 71, 107
chemical reaction, 41
chemical synthesis, 9
cold working, 35
composite, 107
corrosion, 53

ductility, 1

electrical properies, 119
embrittlement, 29

fatigue, 53
Fe, 89, 113, 125
fracture, 137

grain size, 161

kinetics, 77, 101, 125, 137, 161

laser decomposition, 65

magnetic properties, 149
mechanical alloying, 89
metal, 155
microstructure, 125, 131, 149
mixtures, 101
Mo, 113

nanostructure, 71

oxidation, 119
oxide, 59

Pb, 21
phase transformation, 47, 131, 143, 155, 161
phase transformations, 131
polymer, 137

rapid solidification, 149

simulation, 1, 21, 29, 35, 41
sintering, 9
steel, 1, 29, 35, 83, 95, 101, 119, 131, 143
strength, 83, 95, 143, 155
structural, 47, 77
synthesis, 65, 71, 89, 107

Ti, 59

waste management, 77
welding, 95

Printed in the United States
by Baker & Taylor Publisher Services